SpringerBriefs in Applied Sciences and Technology

# PoliMI SpringerBriefs

**Series Editors**

Barbara Pernici, DEIB, Politecnico di Milano, Milano, Italy

Stefano Della Torre, DABC, Politecnico di Milano, Milano, Italy

Bianca M. Colosimo, DMEC, Politecnico di Milano, Milano, Italy

Tiziano Faravelli, DCHEM, Politecnico di Milano, Milano, Italy

Roberto Paolucci, DICA, Politecnico di Milano, Milano, Italy

Silvia Piardi, Design, Politecnico di Milano, Milano, Italy

Gabriele Pasqui ⓘ, DASTU, Politecnico di Milano, Milano, Italy

Springer, in cooperation with Politecnico di Milano, publishes the PoliMI Springer-Briefs, concise summaries of cutting-edge research and practical applications across a wide spectrum of fields. Featuring compact volumes of 50 to 125 (150 as a maximum) pages, the series covers a range of contents from professional to academic in the following research areas carried out at Politecnico:

- Aerospace Engineering
- Bioengineering
- Electrical Engineering
- Energy and Nuclear Science and Technology
- Environmental and Infrastructure Engineering
- Industrial Chemistry and Chemical Engineering
- Information Technology
- Management, Economics and Industrial Engineering
- Materials Engineering
- Mathematical Models and Methods in Engineering
- Mechanical Engineering
- Structural Seismic and Geotechnical Engineering
- Built Environment and Construction Engineering
- Physics
- Design and Technologies
- Urban Planning, Design, and Policy

http://www.polimi.it

Fulvio Re Cecconi · Ania Khodabakhshian ·
Luca Rampini

# Building Tomorrow: Unleashing the Potential of Artificial Intelligence in Construction

**POLITECNICO**
MILANO 1863

Fulvio Re Cecconi
DAEC
Politecnico di Milano
Milan, Italy

Ania Khodabakhshian
DABC
Politecnico di Milano
Milan, Italy

Luca Rampini
DAEC
Politecnico di Milano
Milan, Italy

ISSN 2191-530X ISSN 2191-5318 (electronic)
SpringerBriefs in Applied Sciences and Technology
ISSN 2282-2577 ISSN 2282-2585 (electronic)
PoliMI SpringerBriefs
ISBN 978-3-031-77196-5 ISBN 978-3-031-77197-2 (eBook)
https://doi.org/10.1007/978-3-031-77197-2

# Contents

# Chapter 1
# Is Attention All We Need?

It will not have escaped most people's attention that the question in the title is based on the title of a scientific article that revolutionised the field of Artificial Intelligence (AI). The article, published in 2017 and written by eight scientists working at Google (Vaswani et al. 2017), introduced a new deep learning architecture, known as "Transformers" (Lin et al. 2022), based on the attention mechanism. This architecture revolutionised the field by AI from scientists to the general public, thanks to the General Pre-Trained Transformers (GPT) on which, for example, ChatGPT is based.

This democratisation of AI—you don't need a degree to use it—coupled with the vast capabilities of Large Language Models (LLM), has led more and more people to ask how they can use it in the construction industry, if not run small experiments in their own businesses. Paradoxically, the myriad of tests have been driven more by employees and technicians than by top management, and almost always using their own AI, i.e. their own account on ChatGPT or Gemini, rather than a company's AI. This gave rise to the acronym BYOAI (bring your own artificial intelligence) in the wake of BYOD (bring your own device) (Miller et al. 2012), which emerged when the proliferation of laptops, often more powerful and better equipped than corporate ones, led employees to use personal devices at work. BYOAI, despite its limitations in terms of the security and privacy of corporate data, has greatly facilitated the spread of AI by generating significant research demand even in areas of the construction industry where generative artificial intelligence is not required. This huge demand for research is illustrated in Fig. 1.1, which shows the results of a query on the Scopus database. The query returned over 43,000 results, of which almost 20,000, or almost half, were after 2022, indicating that the impact of ChatGPT, which was introduced in 2022, on research in the construction sector was huge.

Often regarded as one of the cornerstones of modern economic development, the construction industry plays a critical role in shaping the physical, social and economic infrastructure of society (Lopes 2012). Encompassing a wide range of activities from the design and construction of buildings and infrastructure to their maintenance and

F. Re Cecconi et al., *Building Tomorrow: Unleashing the Potential of Artificial Intelligence in Construction*, PoliMI SpringerBriefs, https://doi.org/10.1007/978-3-031-77197-2_1

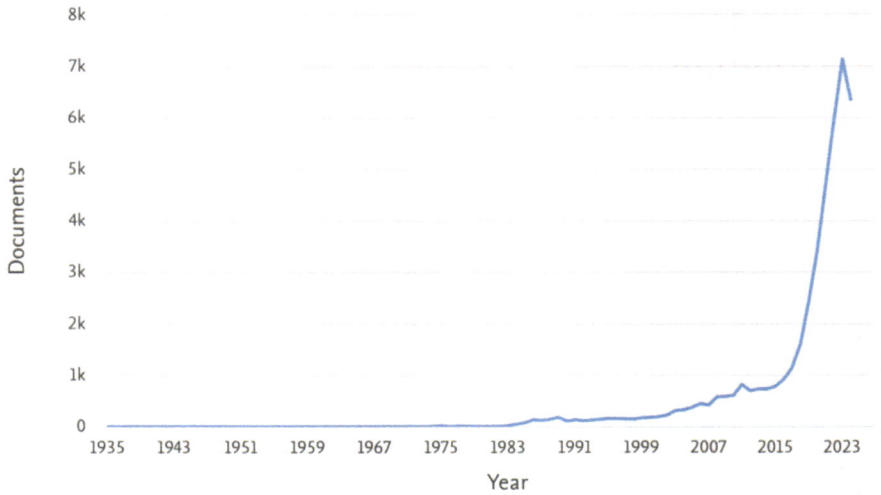

**Fig. 1.1** Number of research works indexed in Scopus related to AI in construction. The drop in 2024 is due to the fact that the year wasn't finished when the search has been performed—search key: (((machine OR deep) AND learning) OR (artificial AND intelligence)) AND construction*—search made on September the 26th 2024

operation, the Architecture, Engineering, Construction and Operation (AECO) sector is central to both urban and rural development. Despite its importance, the construction industry has long been characterised by a slow adoption of new technologies, especially digital innovations, compared to other sectors such as manufacturing or finance (Almatari et al. 2023). Over the past two decades, however, increasing pressure for sustainability, efficiency and innovation has led to a growing interest in integrating advanced technologies, particularly AI, into construction processes (Blanco et al. 2023).

This book explores the intersection of AI and the construction industry, looking at how this transformative technology can address critical challenges in the AECO sector. From improving project management and sustainability to enhancing risk management and operational efficiency, AI promises to revolutionise various aspects of construction. However, its adoption remains fraught with challenges, ranging from technological readiness and data availability to the conservatism of the industry and the complexity of its operating environment. The following chapters of this book offer an in-depth examination of these issues, tracing the trajectory of AI adoption and its potential to reshape the future of construction.

## 1.1 Structure of the Book

This book is divided into several key chapters, each focusing on different aspects of AI's integration into the AECO sector, providing a comprehensive overview of both the opportunities and challenges.

Chapter 2, "Is the Construction Industry Ready for AI?", sets the stage by exploring the current state of AI adoption in the construction industry. It provides a critical analysis of the barriers to AI implementation, such as risk aversion, cost constraints, and the industry's fragmented structure. Through the lens of the Technology-Organisation-Environment (TOE) framework, the chapter introduces the AI Readiness Index (AIRI), a tool developed to assess how prepared AECO firms are to adopt AI technologies. The chapter concludes with insights into the current gap between the industry's enthusiasm for AI and its actual preparedness.

Chapter 3 focuses on the role of AI and Big Data in enhancing sustainability in construction, particularly during the early design stages of projects. Early-stage decisions have long-term implications for the environmental, economic, and social sustainability of construction efforts. By leveraging AI to analyse large datasets, architects and engineers can make informed decisions that optimise material selection, energy use, and operational efficiency. The chapter also explores the potential of AI-driven energy retrofitting, supported by machine learning models, to improve the sustainability of existing buildings.

Chapter 4, "AI and Risk Management in Construction," examines the use of AI and ML techniques in improving risk management processes. Traditionally, risk management in construction has been subjective and reliant on expert judgment, often leading to inefficiencies. AI offers tools such as Neural Networks and Bayesian Networks to predict, analyse, and mitigate risks more effectively. This chapter offers a comparative analysis of probabilistic and deterministic AI models and their application to various aspects of construction risk management, such as safety, project scheduling, and cost estimation.

Chapter 5 shifts the focus to the use of Computer Vision (CV) in asset and facility management. Facility management is crucial in ensuring the long-term performance of buildings and infrastructure. Computer vision technologies, combined with Building Information Modelling (BIM), offer innovative solutions for tasks like 3D reconstruction, condition monitoring, and object detection. However, the chapter highlights challenges such as data scarcity and the need for more efficient methods to create digital twins of existing assets. The potential of synthetic data to supplement real-world datasets in training AI models is also explored.

Chapter 6 addresses the transition from Industry 4.0 to Industry 5.0 within the construction sector. Industry 4.0, also known as Construction 4.0, emphasises the use of digital technologies like robotics, IoT, and AI to optimise construction processes. However, Industry 5.0 shifts the focus towards a more human-centred and ethical

approach, where collaboration between humans and intelligent machines is priori-
tised. This chapter discusses the ethical implications of AI in construction, particularly regarding job displacement, data privacy, and algorithmic bias, while advocating
for Responsible AI frameworks that ensure fairness, transparency, and accountability.

## 1.2   Bridging the Gap Between Potential and Practice

As the construction industry grapples with the integration of AI, a recurring theme
throughout this book is the gap between the technology's potential and its practical implementation. While AI offers significant opportunities to improve the efficiency, sustainability and safety of construction, its adoption is hindered by structural
and cultural barriers within the industry. These include the high stakes involved
in construction projects, the risk-averse nature of the industry, and the lack of
standardised AI solutions tailored to the unique challenges of construction.

To fill this gap, this book not only highlights the technical aspects of AI applications, but also examines the broader organisational, environmental and ethical
considerations that need to be addressed. From AI readiness assessments to responsible AI frameworks, the book provides both industry professionals and researchers
with the tools and insights needed to facilitate AI-driven digital transformation in
the construction sector.

By examining case studies, presenting empirical data, and offering theoretical
frameworks, this book aims to provide a comprehensive guide to understanding the
current landscape of AI in construction. It also serves as a call to action for stakeholders across the industry to invest in innovation, overcome barriers to adoption,
and embrace the future of construction in the digital age.

In conclusion, while the road to AI integration in construction is fraught with
challenges, the benefits are too great to ignore. As AI technologies continue to evolve,
they will increasingly shape the way construction projects are designed, managed
and operated. This book aims to illuminate this path and provide a roadmap for both
industry leaders and policymakers to harness the power of AI for a more efficient,
sustainable and human-centred construction industry.

## References

H.A.Q. Almatari, M. Chan, M.A.N. Masrom, Factors inhibiting the adoption of industrial revolution
    4.0 in Malaysian construction industry. Smart Sustain. Built Environ. (2023). https://doi.org/10.
    1108/SASBE-10-2022-0232
J.L. Blanco, D. Rockhill, A. Sanghvi, A. Torres, *From Start-Up to Scale-Up: Accelerating Growth
    in Construction Technology*. McKinsey & Company (2023)
T. Lin, Y. Wang, X. Liu, X. Qiu, A survey of transformers. AI Open **3**, 111–132 (2022). https://doi.
    org/10.1016/j.aiopen.2022.10.001

J. Lopes, Construction in the economy and its role in socio-economic development, in *New Perspectives on Construction in Developing Countries*. Routledge (2012), pp. 40–71

K.W. Miller, J. Voas, G.F. Hurlburt, BYOD: security and privacy considerations. It Prof. **14**(5), 53–55 (2012)

A. Vaswani, N. Shazeer, N. Parmar, J. Uszkoreit, L. Jones, A.N. Gomez, Ł. Kaiser, I. Polosukhin, Attention is all you need, in *Advances in Neural Information Processing Systems*, December 2017 (2017)

# Chapter 2
# Is the Construction Industry Ready for AI?

Recent advances in Artificial Intelligence (AI) have offered the Architecture, Engineering & Construction (AEC) sector unprecedented automation opportunities (Darko et al. 2020). Attracted by AI potential, many businesses have begun planning how to make the necessary transition (Ferrigno et al. 2023; Presti et al. 2023). There is a rapid growth of tech investment in the AEC sector, including the efforts of adopting some kind of AI capability (Blanco et al. 2023).

Nevertheless, despite the enthusiasm surrounding the use of AI, 85% of current big data projects fail (Satell 2018). As a consequence, before embarking on the path to AI, practitioners must have a clear understanding of the problems that AI should solve and what is the necessary foundation to ensure a smooth implementation of the adoption plan (Agrawal et al. n.d.).

Emerging technologies, including AI, serve as the foundation for solving the world's most challenging problems of the twenty-first century (Groen and Walsh 2013). Given the potential benefits of new technologies, significant resources are invested in Research & Development (R&D) (Coccia 2019; Neirotti and Pesce 2019), and engineering managers carefully create procedures to integrate them. Attempts to incorporate new technologies into processes before they are mature have been demonstrated to increase project costs and schedules (Katz et al. 2015).

But, how to determine when a technology is ready to go from research and development to commercial or system development? How to assess the advancement of technological readiness? When confronted with these issues in the 1970s, engineers at the National Aeronautics and Space Administration (NASA) devised the Technological Readiness Level (TRL) scale (Mankins 1995; Sadin et al. 1989). The TRL scale has gone outside NASA since its beginnings, and it is currently used not just on aerospace and defense projects, but also on projects in industries such as energy, transportation, and electronics. Despite this increased use, technology maturity evaluation is a difficult task; hence, other models, such as Gartner's Hype Cycle, the

By Fulvio Re Cecconi and Luca Rampini.

© The Author(s), under exclusive license to Springer Nature Switzerland AG 2024
F. Re Cecconi et al., *Building Tomorrow: Unleashing the Potential of Artificial Intelligence in Construction*, PoliMI SpringerBriefs,
https://doi.org/10.1007/978-3-031-77197-2_2

software Capability Maturity Model, and other data-driven techniques, have been created (Albert et al. 2015; Dedehayir and Steinert 2016; Humphrey 1988).

## 2.1  A Traditionally Backward and Conservative Industry

Several factors contribute to the perception that construction industry is traditionally backward and more conservative if compared to other industries. The first one of these factors is risk aversion (Kim and Reinschmidt 2011). Construction projects require significant financial investments and entail potential risks. The industry has traditionally been risk-averse, as failures can have severe consequences. This risk aversion can lead to resistance to adopting new technologies or methodologies, as the industry tends to favour proven and well-established practices.

The fragmented nature of the construction industry (Toland and Collery 2023) is another significant obstacle to innovation. The industry's supply chain is often fragmented, with multiple stakeholders, including architects, engineers, contractors, subcontractors, and suppliers. This fragmentation can lead to communication challenges and a reluctance to embrace change due to the coordination required among diverse parties. Moreover, the construction industry is often perceived as backward and conservative due to factors such as regulatory compliance and safety concerns (Almatari et al. 2023). Strict regulations and safety standards are in place, which require adherence to established processes and practices. However, this emphasis on safety and compliance can sometimes hinder innovation, as new methods or technologies may be seen as unproven or risky.

The extended lifecycle of construction projects is a significant obstacle to innovation in the industry (Akintoye et al. 2012). Due to the long duration of projects, from design to end of life, it can be difficult for the industry to keep up with rapidly evolving technologies and methodologies. This can lead to a conservative mindset, as new approaches may disrupt ongoing projects. To overcome this challenge, the industry must be open to adopting new approaches and technologies.

A fifth factor contributing to the lack of innovation in the construction sector, compared to other industries, is cost and budget constraints (Shojaei and Burgess 2022). The construction industry is highly cost-sensitive, and profit margins can be slim. The pressure to control costs and adhere to budget constraints may discourage experimentation with new technologies or methods that could be perceived as expensive or unproven.

The absence of standardization is a hindrance to innovation in the construction industry (Shojaei and Burgess 2022). Unlike some industries that benefit from standardized processes and components, construction projects are often unique and tailored to specific requirements. This lack of standardized practices can make it more challenging to implement widespread changes or innovations across the industry.

Eventually, innovative practices may face resistance due to the inertia and tradition prevalent in the construction industry (Wuni et al. 2024). Traditional methods, passed down through generations, have been perceived as successful in the past, making it challenging to break away from established practices.

Despite these challenges, there is a growing recognition within the construction industry of the need for innovation and adaptation. Efforts are being made to integrate digital technologies, embrace sustainable practices, and improve overall project management.

Researchers have studied the application of AI in AEC specific problems during the last few decades (Darko et al. 2020; Osei and Cheng 2024; Pan and Zhang 2021; Rampini and Re Cecconi 2022). Despite encouraging studies, AI use in the industry is limited due to different issues, including data scarcity and lack of people with the necessary skills (Agrawal et al. n.d.; Rampini et al. 2022). The variety of possible causes hindering AI adoption suggests that each company must pursue a strategy to overcome them. Therefore, they need a methodology to assess their readiness and plan accordingly.

## 2.2  What is a Readiness Metric?

How does one determine when a technology is ready to go from research and development to commercial or system development? How can one assess the advancement of technological readiness? Faced with these problems in the 1970s, engineers at the National Aeronautics and Space Administration (NASA) developed the TRL scale (Mankins 1995; Sadin et al. 1989). Since its inception, the TRL scale has spread beyond NASA and is now used not only for aerospace and defence projects, but also for projects in industries such as energy, transportation and electronics. Despite this increased use, assessing technology maturity is a difficult task, which has led to the creation of other models such as Gartner's Hype Cycle, the Software Capability Maturity Model, and other data-driven techniques (Albert et al. 2015; Dedehayir and Steinert 2016; Humphrey 1988).

In the Information and Communications Technology (ICT) field, the term AI readiness refers to "an organization's abilities to deploy and use AI in ways that add value to the organization" (Holmström 2022); thus, AI readiness entails more than just technology. This work presents a new, tailored metric for analysing the AI readiness of AEC firms, taking into account an evaluation framework based on the Technology-Organisation-Environment (TOE) methodology. The TOE framework is a multi-perspective theory established to provide a method for examining the adoption of Internet services at the firm level (Tornatzky et al. 1990). In this study, the framework identified seven key dimensions: vision, data, system, digital foundation, talent, value, and governance. These seven dimensions were assessed using a survey targeted at a group of small, medium and large companies that could reflect the size distribution of the industry, which is characterised by a significant presence of small and medium enterprises (SMEs). Moreover, by analysing the results, based on the firm's size, it is possible to understand where the drivers for AI adoption in AEC might be. The findings are finally synthesized in a simple and communicative parallel axis plot that should guide the company's investment decision in future R&D to implement AI technologies. Overall, the framework has a twofold function: to help companies understand which dimensions they need to focus on if they want to introduce AI and help governments and experts to understand the AI readiness of the industry, which is a critical area in the path toward digitalization.

In the next section, the chapter provides details about the theoretical background used for developing the metric. This is followed by a deeper description of the AI Readiness Index for AEC and the survey conducted to validate the utility of the metric for companies. The chapter is concluded with the discussion of the findings and their implications for the AEC industry.

## 2.3  AI Readiness Metric for the Construction Industry

The Artificial Intelligence Readiness Index (AIRI) metric is created using the TOE framework. As the name implies, the TOE framework examines a company from three areas: technology, organization, and environment. At the organizational level, the TOE framework explains factors that influence adoption decisions. (Tornatzky et al. 1990) discovered that technological factors and organizational and environmental contexts influence the decision to implement innovation at the firm level.

### 2.3.1  TOE Areas

The TOE framework key concept is that parts of technological, organizational, and environmental factors are required for AI adoption. The research hypotheses are presented from perspectives on technology preparedness, organizational preparation, and environmental readiness. The TOE framework, unlike other adoption theories, does not provide a collection of characteristics that influence innovation uptake (Aboelmaged 2014). As a result, the factors we have chosen are assumptions based on the literature and experience from the AEC field. The TOE framework is based on the analysis of three principal areas:

- Technological readiness: The ability of a company to absorb new technology is referred to as technological readiness (Richey et al. 2007). This encompasses internal (technology infrastructure) and external (market-available) technologies relevant to the firm. The organization must properly consider the difficulties and challenges of adopting new technology. As a result, technological readiness offers a better way to predict the benefits gained from technological adoption (Richey et al. 2007).
- Organizational readiness: (Iacovou et al. 1995) define organization readiness "as the availability of the needed organizational resources for adoption." Based on

previous studies of innovation (Yang et al. 2015), organizational parameters such as business size, top management, and so on influence the adoption of innovations.

- Environmental readiness: In general, organizations execute their operations in relation to their surroundings. This involves how a company does business with its competitors in the same industry. The term "environmental preparedness" relates to how an organization perceives external factors influencing its decision to implement AI. According to research, external factors such as competitive pressure and regulatory difficulties are driving factors in adopting new innovation (Ifinedo 2005).

The three areas guided the definition of the dimensions that will formulate the AIRI. These dimensions are key aspects to consider when evaluating AI readiness in AEC industry. The dimensions were identified according to the three areas of the TOE framework and by analysing the literature and industry's expertise. In particular, the seven dimensions are:

- Vision: How firms should ensure that AI transformations are aligned with strategic goals in order to have a meaningful business impact;
- Data: How companies acquire the right data and how they are accessible and transferable among the company's processes;
- System: How firms choose the correct technology stack to support the construction and use of AI-powered analytical applications from start to finish;
- Digital foundation: What is the existing level of the company in using digital technologies;
- Talent: How firms address the skill transition challenge: recruiting, training, and establishing AI communities;
- Value: How firms plan to build durable, AI-powered assets to maximize their impact.
- Governance: How firms plan to scale AI across all their business processes.

Specific and tailored questions were formulated to assess companies' performance in each dimension (Table 2.1).

## 2.3.2  AI Readiness Index

Companies' strategic decisions on whether and where to invest are frequently based on subjective judgments or experiences that may yield poor outcomes (Liu and Jiang 2012). Therefore, a more scientific approach that helps companies evaluate their AI readiness and direct their investments towards the right dimensions is the objective for introducing a new, tailored metric: the AIRI.

The evaluation parameters are based on the seven dimensions that were previously introduced. For business entities that have already adopted AI, AIRI will also assess the gap between the current and desired state, thus enabling organizations to implement programs to achieve maximum efficiency.

**Table 2.1**  Proposed questions to assess the AI readiness dimensions

| Question | Dimension | Area |
| --- | --- | --- |
| What best describes your company's AI strategy? | Vision | Organizational |
| What best describes your organisation's approach to data acquisition? | Data | Organizational |
| How does your organization acquire data and ensure data quality? | Data | Organizational |
| What best describes your organization's AI tools? | System | Technology |
| What percentage of AI projects (compared to the total) have been integrated into business processes)? | System | Technology |
| Are there specific, customised tools for data analysis in your organization? | System | Technology |
| What other technological innovations, excluding AI, has your company adopted in recent years? | Digital foundation | Organizational/ Technology |
| What best describes staff involvement regarding AI? | Talent | Organizational |
| What is your organization's approach regarding AI value? | Value | Environmental |
| What best describes your organization's approach to AI governance? | Governance | Environmental |

Ultimately, AIRI tries to translate abstract concepts into concrete actions to help organizations accelerate AI adoption. Before going into the details of describing AIRI, it is appropriate to establish a vocabulary of terms used to avoid misunderstandings. The vocabulary is given in the Table 2.2.

**Table 2.2**  AIRI assessment table

|  | AI unaware | AI aware | AI ready | AI competent |
| --- | --- | --- | --- | --- |
| AIRI | −100 to −50% | −50–0% | 0–50% | 50–100% |
| General capanilities | Might hear about AI but is unaware of applications | Potential consumers of AI solutions. Capable of identifying use cases for AI application | Capable of integrating pre-trained AI model into AECO processes | Capable of developing customised AI solutions for specific business needs |
| General Characteristics | Wait for vendors to convince use cases and business value of AI | Identified potential use cases and seek AI solutions from vendors | Evaluated viability of pre-trained models | Developed roadmap for AI implementation |
| AI Adoption Suitability | Consume ready-made end-to-end AI solutions | Consume ready-made end-to-end AI solutions | Integrate pre-trained AI models and solutions for common AI applications | Developed customised AI model for unique business needs |

Given these seven dimensions, the AIRI is computed according to the following equation:

$$AIRI = s \times nd_p + w \times nd_n/7 \qquad (2.1)$$

where:

$s$ = the AI readiness strengths index.

$w$ = the AI readiness weaknesses index.

$nd_p$ = the number of dimensions with a positive score.

$nd_n$ = the number of dimensions with a negative score.

The terms included in the AIRI metric are thoroughly explained in the next paragraphs.

*Strength and Weakness Indexes*

Strength and Weakness Indexes give a preliminary assessment of the dimensions where companies performed well or poorly, respectively. Those indexes' value depends on the scores and the number of dimensions (good or bad). The first step in calculating AIRI is collecting the company's evaluation on each of the seven dimensions, obtained as follows (Eq. 2.2):

$$q_d = \sum_{i=1}^{I} \frac{a_{d,i}}{I} \qquad (2.2)$$

where:

$a_{d,i}$ = the evaluation of the $i$th question of the $d$ dimension.

$I$ = the total number of questions in the $d$ dimension.

$q_d$ = the AI readiness score for dimension $d$.

In the definition of AIRI (Eq. 2.1), the positive mark is computed from the positive scores according to the Eq. 2.3:

$$nd_p = \sum_{d=1}^{7} \begin{cases} 1 \, if p_d \geq 0.5 \\ 0 \, if p_d < 0.5 \end{cases} \qquad (2.3)$$

where:

$p_d$ = the AI readiness score of the d dimension.

Similarly, the negative mark is computed from the negative scores according to the Eq. 2.4:

$$nd_n = \sum_{d=1}^{7} \begin{cases} 0 \, if p_d \geq 0.5 \\ 1 \, if p_d < 0.5 \end{cases} \qquad (2.4)$$

The AI readiness score of the d-th dimension $p_d$ is computed from the evaluation of the d-th dimension $q_d$ according to the following Eq. 2.5:

$$p_d = (q_d - 1)/4 \qquad (2.5)$$

The AI readiness strengths index $s$ is an evaluation of all the positive aspects measured through the survey. It is a number ranging from 0 to 1 computed according to the following Eq. 2.6:

$$s = s' - nd_p \times 0.5/3.5 \qquad (2.6)$$

where:

$s'$ = the sum of the dimensions where the score is higher than 0.5 in the following Eq. 2.7:

$$s' = \sum_{d=1}^{7} \begin{cases} p_d \, if p_d \geq 0.5 \\ 0 \, if p_d < 0.5 \end{cases} \qquad (2.7)$$

The AI readiness weaknesses index $w$ is computed in an similar way and assesses the negative aspects highlighted by the survey.

$$w' = \sum_{d=1}^{7} \begin{cases} 0 \, if p_d \geq 0.5 \\ p_d \, if p_d < 0.5 \end{cases} \qquad (2.8)$$

$$w = \left(w' - nd_n \times 0.5\right)/3.5 \qquad (2.9)$$

To make the AIRI metric accessible, it is essential to represent its information through data visualization principles. Data visualization is the graphical representation of information and data (Chen et al. 2008). Furthermore, it allows workers or business owners to deliver facts to non-technical audiences without causing misunderstanding. Therefore, in this study, the AIRI is represented in a dashboard like Fig. 2.1. Each company receives a final dashboard at the end of the survey and can interpret the results immediately. The dashboard visualization condensates all the relevant information graphically, in particular:

- How the company performed in each dimension (value from 0 to 1). These scores allow companies to immediately understand which dimensions are already at a level consonant with AI and which ones need more investment;
- The number of positive or negative dimensions;
- Strength and Weakness Index (positive and negative mark);
- The evaluation of AIRI as percentage (from −100 to 100%).

The companies are classified into four categories depending on the AIRI percentage. The meaning of each category is summarized in Table 2.3. The AIRI metric guided the assessment of the AI-readiness of AEC sector obtained by surveying companies and practitioners.

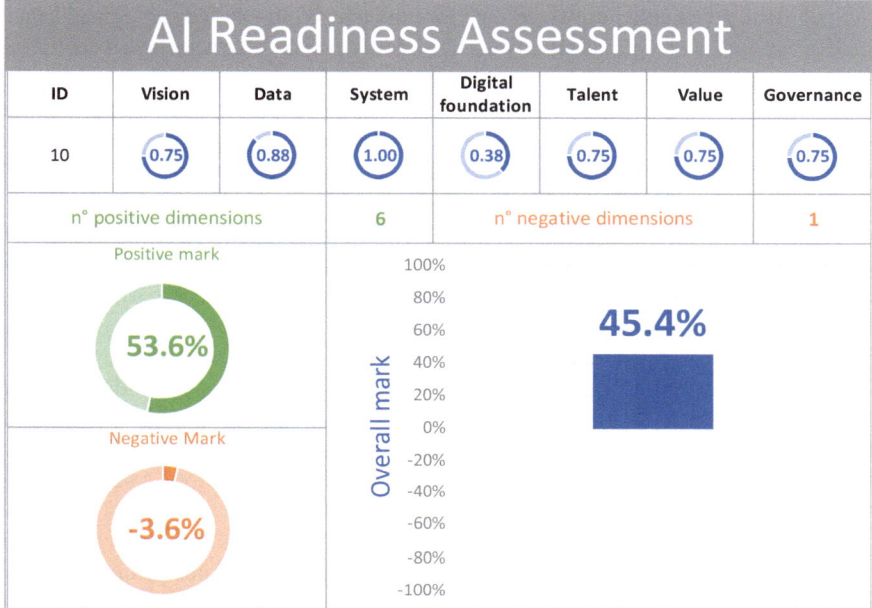

**Fig. 2.1**  AI readiness dashboard

**Table 2.3**  Average evaluations for each dimension

| Size | Vison | Data | System | Digital foundation | Talent | Value | Governance |
|------|-------|------|--------|--------------------|--------|-------|-----------|
| All  | 1.87  | 2.42 | 1.88   | 2.08               | 1.97   | 2.27  | 1.57      |
| >500 | 1.83  | 2.58 | 1.79   | 2.25               | 1.92   | 2.58  | 1.67      |
| <500 | 1.89  | 2.31 | 1.94   | 1.97               | 2.00   | 2.06  | 1.50      |
| <20  | 2.14  | 2.43 | 2.07   | 2.00               | 2.00   | 2.14  | 1.43      |

## 2.3.3   AIRI Validation Through Survey

There are numerous methodologies available for collecting data on technology adoption and policy opinions. However, the complex issue of AI integration in the AEC industry cannot be studied solely in a quantitative manner; it necessitates the use of qualitative research methodologies. Traditional surveys, interviews, and panel discussions are a few of the relevant and widely used methods for this type of research. Each method has inherent advantages and disadvantages that must be carefully considered before making a decision. To encompass a wider range of industry professionals, we conducted online surveys to foster a deep understanding of the problem while producing the most useful data set for analysis and utilization.

Specifically, this study collected data through a multiple-choice survey and Likert scale questions: the former is mainly used to understand the characteristic distribution of the respondents, while the latter is used to evaluate the companies' level of AI readiness. The evaluation is performed by analysing seven dimensions through the questions proposed in Table 2.1, identified by combining information from previous research (Rampini et al. 2022) and relevant literature (*AI Singapore—AIRI* 2022; Dataiku—AI Maturity Survey 2022).

The data was collected via a Microsoft Forms survey, using its online format to distribute it across various platforms and social media sites such as LinkedIn and Twitter. The survey is divided into two main parts. The first part, consisting of multiple choice questions, collects information about the respondents, including their company size, revenue and role. Despite the anonymity of the survey, the data proved helpful in further examining the level of preparedness in the industry, from large corporations to tech start-ups. A detailed five-point Likert scale was used in the second part to assess the status of the industry in each dimension. In this study, the minimum score for a dimension to be considered ready to apply AI is 3: a score below 3 means that more efforts need to be made in that specific dimension before AI can be implemented. On the other hand, a score above 3 means that the dimension is already prepared and developed for inclusion in an AI process. These sections helped to assess the current level of readiness in the industry, providing insight into which dimensions companies need to focus more on if they want to make the transition to AI-driven processes.

### 2.3.4  Analysis of the Panellists

The analysis of the respondents helps to put the results in context and to check whether the distribution of panellists reflects the panorama of the AEC sector, which is mainly composed of SMEs, but strongly influenced by the large companies working on it. The size of the respondents is shown in Fig. 2.2, together with the stages of the asset lifecycle in which these companies operate.

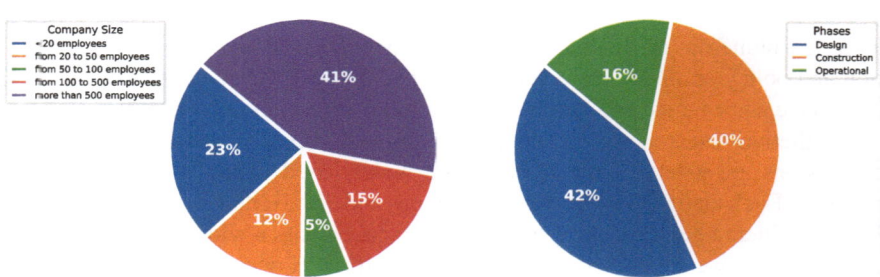

**Fig. 2.2** Size of the respondents, together with the stages of the asset lifecycle in which these companies operate

The distribution of panellists is in line with the industry market, particularly in Italy where most of the responses come from. A large proportion of respondents work in large companies with more than 500 employees (42%), while the remainder work in SMEs (58%), particularly in companies with less than 20 employees (23%). In addition, most respondents work in the design and engineering phase (84%).

## 2.3.5 Survey's Results

The results presented in this paragraph are derived from the second section of the survey, i.e., the one linked with evaluating the company's readiness in the seven dimensions through a Likert scale from 1 to 5. The evaluations of the seven AI readiness dimensions are presented in Table 2.3 as the average of all results. Furthermore, the results are also broken down by large, medium, and small companies to better understand the influence of company size on AI readiness.

The average of many scores sometimes conceals interesting results; therefore, a more detailed analysis may be appropriate. In cases such as Likert scale questionnaires, this is crucial because an average evaluation might not be representative of the evaluation's distribution since the average value is seldom one of the grades assigned in the questionnaire. Analysing the survey responses of all the companies collectively in Fig. 2.3a, the majority of answers are concentrated in evaluation 1 (63% for dimension 7; 50% for 1; 47% for dimensions 3 and 5) with the exception of dimensions 4 and 6 where the majority of answers, 40 and 47% respectively, are in score 2. If, on the other hand, one analyses the answers to the questionnaire by company size, one can see that in large companies, the number of answers equal to 2 is closer to the number of answers equal to 1 than in the cases of small or micro companies. In two cases, for sizes 2 and 4, the answers of the large companies were higher (50%) on evaluations 4 and 3.

In statistics, a correlation analysis determines the degree to which a pair of variables are linearly related. The analysis might help in establishing the relationship between variables, whether causal or not. The correlation matrix in Table 2.4 provides the Pearson coefficient ($\rho$) for all data collected with the questionnaire. In some cases, the overall figure can be misleading, hiding a solid correlation between data subsets based on company size. The low correlation coefficient between the dimensions "System" and "Digital Foundation" shown in Table 2.4, for example, hides a high correlation between the answers given for these two dimensions by companies with more than 500 employees and a very low correlation for those with less than 20. For the former, the correlation coefficient is 0.707, and for the latter, 0.338 (Fig. 2.4a).

In other cases, the company's size heavily influences the correlation coefficient of two AI readiness dimensions. For example, in Fig. 2.4b, it can be observed that against an overall $\rho = 0.569$ between the dimensions "Vision" and "Data" the Pearson's correlation coefficient for the three sizes of companies is: $\rho = 0.703$ for large companies (>500 Employees), $\rho = 0.648$ for SMEs (<500 Employees) and $\rho = 0.926$ for small companies (<20 Employees).

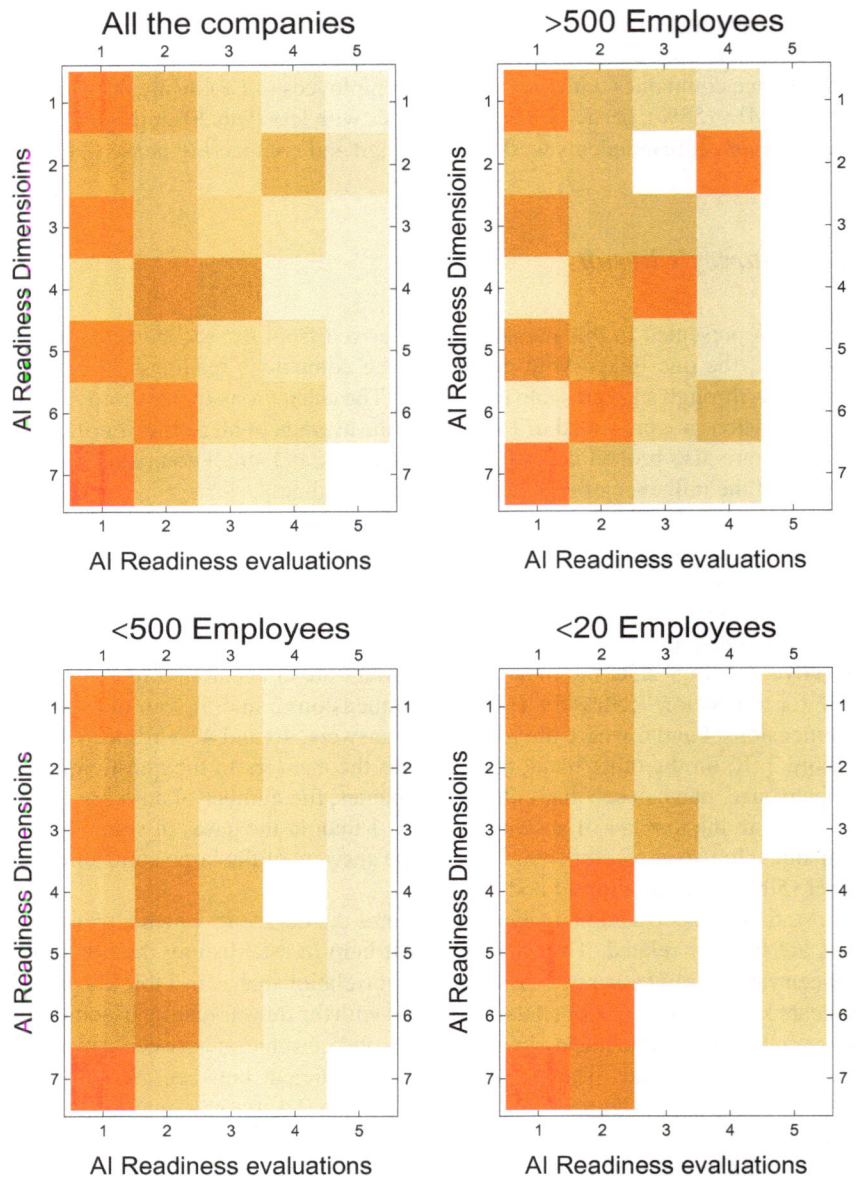

**Fig. 2.3** AI readiness evaluations distributions for different company's size

**Table 2.4**  Correlation matrix in terms of Pearson coefficient

|                   | Vision | Data | System | Digital foundation | Talent | Value | Governance |
|-------------------|--------|------|--------|--------------------|--------|-------|------------|
| Vision            | 1      | 0.66 | 0.79   | 0.76               | 0.77   | 0.49  | 0.73       |
| Data              | 0.66   | 1    | 0.55   | 0.59               | 0.54   | 0.39  | 0.54       |
| System            | 0.79   | 0.55 | 1      | 0.53               | 0.88   | 0.63  | 0.81       |
| Digital foundation| 0.76   | 0.59 | 0.53   | 1                  | 0.54   | 0.33  | 0.59       |
| Talent            | 0.77   | 0.54 | 0.88   | 0.54               | 1      | 0.5   | 0.71       |
| Value             | 0.49   | 0.39 | 0.63   | 0.33               | 0.5    | 1     | 0.51       |
| Governance        | 0.73   | 0.54 | 0.81   | 0.59               | 0.71   | 0.51  | 1          |

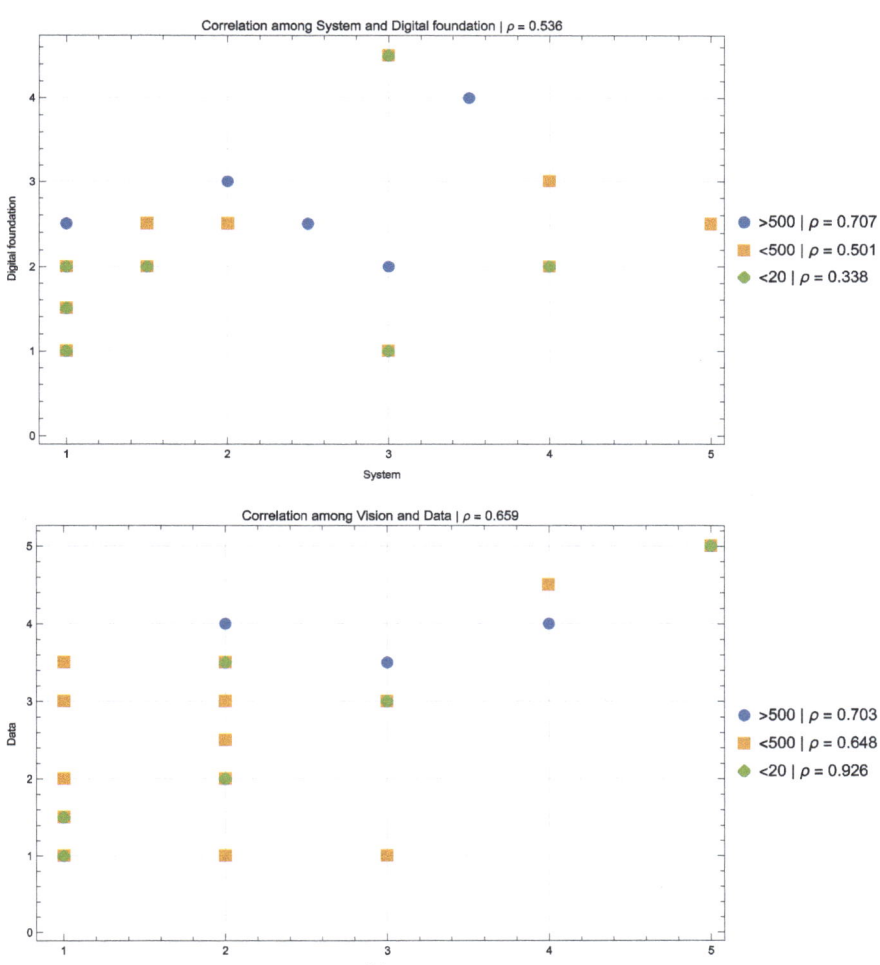

**Fig. 2.4**  Examples of dimension's correlation

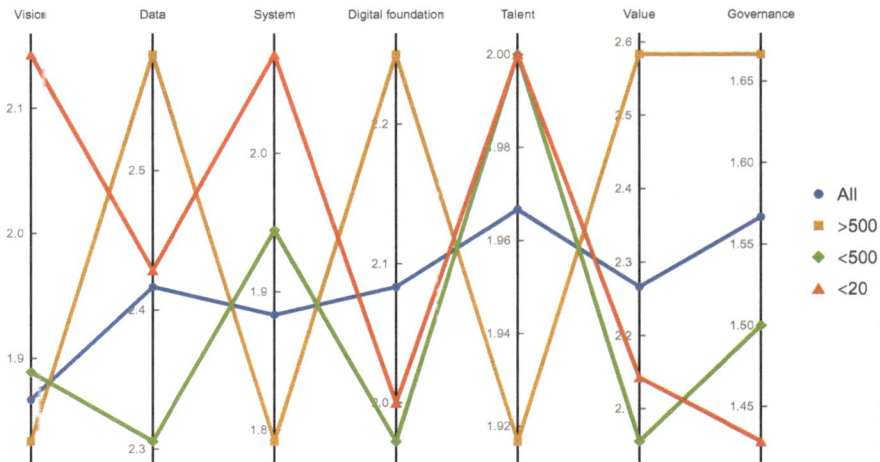

**Fig. 2.5** Parallel axis plot showing the readiness of the industry (blue line) and different subset of companies. The plot can also be used by a single company to compare its AI readiness to the one of its competitors

Based on the survey's results, a parallel axis plot was created to help AI adopters easily evaluate their demands and assess them in each readiness dimension for continued development toward successful industrial applications Fig. 2.5. The plot is a readable tool to compare the AI readiness of either single firms or companies subsets.

Large companies with 500 or more employees scored better on the questionnaire. As can be seen in Fig. 2.5 the line representing them (yellow ochre) is above the line of the average of all responses (blue) in four out of seven dimensions. This result was quite predictable; it is surprising, however, that there are three dimensions where large companies perform worse than average. Specifically, those dimensions are "Vision", "System" and "Talent". One possible explanation for this is that responders from large corporations are more conscious of their constraints. Being employed in a large company that has already started the AI adoption process, they are more aware of the company's limitations in systems and talents employed for AI. They probably see AI as an exploratory field without setting a clear vision strategy for implementing it. On the other hand, unexpectedly, small companies (i.e., with less than 20 employees) generally per- form better than medium enterprises, and in three dimensions out of seven, have the best performance. This result can be explained by emphasizing that most small companies are innovation-oriented startups that place great emphasis on new digital technologies.

Overall, the AECO sector is still quite far from embracing AI processes. The average value in each dimension is below 3, which is the minimum value to reach to implement AI strategies and processes successfully. The industry needs more research and case studies that stress the potential and the framework to adopt for

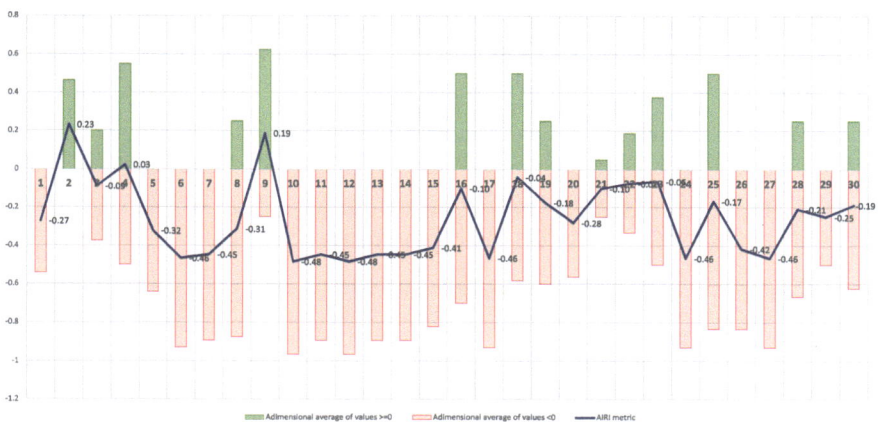

**Fig. 2.6** AI Readiness performance of the thirty surveyed companies

introducing AI. This evaluation tool might help to understand in which dimension we need to focus more in order to facilitate AI adoption.

Finally, the aggregation of the overall results using the AIRI dashboard can give a general picture of the AECO sector status in adopting AI solutions. Figure 2.6 shows all the marks collected during the survey: The green histograms represent the Strength Index, the red histograms represent the Weakness Index, and the overall AIRI metric is represented by the blue line. The results showed in Fig. 2.6 confirmed the insights from the previous analysis: many companies in the AECO industry do not reach the minimum level of AI readiness. The few companies that registered a positive score in almost all dimensions (there is only one company with all positive scores) are startups.

This result differs from what has been found in other areas (Bessen et al. 2018). Probably, the fragmented, hierarchical economy of the AECO industry hinders disruptive innovations in big firms that could undermine interoperability with SMEs (Barbosa et al. 2017). On the other hand, flexible and innovative startups are more likely to come up with innovative solutions to stand out in the marketplace, as confirmed by the amount of venture capital invested in US Construction Tech that annually has risen from around $250 million in 2013 to well over $1 billion in 2018 (Barbosa et al. 2017).

## References

M.G. Aboelmaged, Predicting e-readiness at firm-level: an analysis of technological, organizational and environmental (TOE) effects on e-maintenance readiness in manufacturing firms. Int. J. Inf. Manag. **34**(5), 639–651 (2014). https://doi.org/10.1016/j.ijinfomgt.2014.05.002

A. Agrawal, V. Singh, R. Thiel, M. Pillsbury, H. Knoll, J. Puckett, M. Fischer, Digital twin in practice: emergent insights from an ethnographic-action research study, in *Construction Research Congress 2022* (n.d.), pp. 1253–1260. https://doi.org/10.1061/9780784483961.131

AI Singapore—AIRI (2022), https://aisingapore.org/innovation/airi/

A. Akintoye, J.S., Goulding, G. Zawdie, Construction innovation and process improvement, in *Construction Innovation and Process Improvement* (2012). https://doi.org/10.1002/9781118280294.ch1

T. Albert, M.G. Moehrle, S. Meyer, Technology maturity assessment based on blog analysis. Technol. Forecast. Soc. Change **92**, 196–209 (2015). https://doi.org/10.1016/j.techfore.2014.08.011

H.A.Q. Almatari, M. Chan, M.A.N. Masrom, Factors inhibiting the adoption of industrial revolution 4.0 in Malaysian construction industry. Smart Sustain. Built Environ. (2023). https://doi.org/10.1108/SASBE-10-2022-0232

F. Barbosa, J. Woetzel, J. Mischke, Reinventing construction: a route of higher productivity (2017)

J.E. Bessen, S. Impink, R. Seamans, L. Reichensperger, The business of AI startups. SSRN Electron. J. (2018). https://doi.org/10.2139/ssrn.3293275

J.L. Blanco, D. Rockhill, A. Sanghvi, A. Torres, From Start-Up to Scale-Up: Accelerating Growth in Construction Technology. McKinsey & Company (2023)

C. Chen, W. Härdle, A. Unwin, *Handbook of Data Visualization* (Springer, Berlin, 2008). https://doi.org/10.1007/978-3-540-33037-0

M. Coccia, Why do nations produce science advances and new technology? Technol. Soc. **59**, 101124 (2019). https://doi.org/10.1016/j.techsoc.2019.03.007

A. Darko, A.P.C. Chan, M.A. Adabre, D.J. Edwards, M.R. Hosseini, E.E. Ameyaw, Artificial intelligence in the AEC industry: scientometric analysis and visualization of research activities. Autom. Constr. **112**, 103081 (2020). https://doi.org/10.1016/j.autcon.2020.103081

Dataiku—AI maturity survey (2022). https://maturity.dataiku.com/survey/

O. Dedehayir, M. Steinert, The hype cycle model: a review and future directions. Technol. Forecast. Soc. Change **108**, 28–41 (2016). https://doi.org/10.1016/j.techfore.2016.04.005

G. Ferrigno, N. Del Sarto, A. Piccaluga, A. Baroncelli, Industry 4.0 base technologies and business models: a bibliometric analysis. Eur. J. Innov. Manag. **26**(7), 502–526 (2023). https://doi.org/10.1108/EJIM-02-2023-0107

A.J. Groen, S.T. Walsh, Introduction to the field of emerging technology management. Creat. Innov. Manag. **22**(1), 1–5 (2013). https://doi.org/10.1111/caim.12019

J. Holmström, From AI to digital transformation: the AI readiness framework. Bus. Horiz. **65**(3), 329–339 (2022). https://doi.org/10.1016/j.bushor.2021.03.006

W.S. Humphrey, Characterizing the software process: a maturity frame-work. IEEE Softw. **5**(2), 73–79 (1988). https://doi.org/10.1109/52.2014

C.L. Iacovou, I. Benbasat, A.S. Dexter, Electronic data interchange and small organizations: adoption and impact of technology. MIS Q. **19**(4), 465–485 (1995). https://doi.org/10.2307/249629

P. Ifinedo, Measuring Africa's e-readiness in the global networked economy: a nine-country data analysis. Int. J. Educ. Dev. Inf. Commun. Technol. (IJEDICT) **1**, 53–71 (2005)

D.R. Katz, S. Sarkani, T. Mazzuchi, E.H. Conrow, The relationship of technology and design maturity to DoD weapon system cost change and schedule change during engineering and manufacturing development. Syst. Eng. **18**(1), 1–15 (2015). https://doi.org/10.1111/sys.21281

H.-J. Kim, K.F. Reinschmidt, Effects of contractors' risk attitude on competition in construction. J. Constr. Eng. Manag. **137**(4) (2011). https://doi.org/10.1061/(asce)co.1943-7862.0000284

Y. Liu, I. Jiang, Influence of investor subjective judgments in investment decision-making. Int. Rev. Econ. & Financ. **24**, 129–142 (2012). https://doi.org/10.1016/j.iref.2012.01.002

J.C. Mankins, Technology readiness levels (1995), http://www.artemisinnovation.com/images/TRL_White_Paper_2004-Edited.pdf

P. Neirotti, D. Pesce, ICT-based innovation and its competitive out-come: the role of information intensity. Eur. J. Innov. Manag. **22**(2), 383–404 (2019). https://doi.org/10.1108/EJIM-02-2018-0039

B.A. Osei, M. Cheng, Preferences and challenges towards the adoption of the fourth industrial revolution technologies by hotels: a multilevel concurrent mixed approach. Eur. J. Innov. Manag. **27**(6), 1912–1937 (2024). https://doi.org/10.1108/EJIM-09-2022-0529

Y. Pan, L. Zhang, Roles of artificial intelligence in construction engineering and management: a critical review and future trends. Autom. Constr. **122**, 103517 (2021). https://doi.org/10.1016/j.autcon.2020.103517

C. Presti, F. De Santis, F. Bernini, Value co-creation via machine learning from a configuration theory perspective. Eur. J. Innov. Manag. **26**(7), 449–477 (2023). https://doi.org/10.1108/EJIM-01-2023-0104

L. Rampini, A. Khodabakhshian, F. Re Cecconi, Artificial intelligence feasibility in construction industry (2022). https://doi.org/10.35490/EC3.2022.189

L. Rampini, F. Re Cecconi, Artificial intelligence in construction as-set management: a review of present status, challenges and future opportunities. J. Inf. Technol. Constr. **27**, 884–913 (2022). https://doi.org/10.36680/j.itcon.2022.043

R.G. Richey, P.J. Daugherty, A.S. Roath, Firm technological readiness and complementarity: capabilities impacting logistic service competency and performance. J. Bus. Logist. **28**(1), 195–228 (2007). https://doi.org/10.1002/j.2158-1592.2007.tb00237.x

S.R. Sadin, F.P. Povinelli, R. Rosen, The NASA technology push towards future space mission systems. Acta Astronaut. **20**, 73–77 (1989). https://doi.org/10.1016/0094-5765(89)90054-4

G. Satell, How to make an AI project more likely to succeed. Harv. Bus. Rev. (2018)

R.S. Shojaei, G. Burgess, Non-technical inhibitors: exploring the adoption of digital innovation in the UK construction industry. Technol. Forecast. Soc. Change **185**, 122036 (2022). https://doi.org/10.1016/j.techfore.2022.122036

P. Toland, D. Collery, How can the Irish construction industry become less fragmented and more productive? in *Proceedings of the European Conference on Knowledge Management, ECKM*, vol. 2 (2023). https://doi.org/10.34190/eckm.24.2.1387

L.G. Tornatzky, M. Fleischer, A.K. Chakrabarti, *The Processes of Technological Innovation*. Lexington Books (1990). https://books.google.it/books?id=EotRAAAAMAAJ

I.Y. Wuni, D.A. Abankwa, K. Koc, S.E. Adukpo, M.F. Antwi-Afari, Critical barriers to the adoption of integrated digital delivery in the construction industry. J. Build. Eng. **83**, 108474 (2024). https://doi.org/10.1016/j.jobe.2024.108474

Z. Yang, J. Sun, Y. Zhang, Y. Wang, Understanding SaaS adoption from the perspective of organizational users: a tripod readiness model. Comput. Hum. Behav. **45**, 254–264 (2015). https://doi.org/10.1016/j.chb.2014.12.022

# Chapter 3
# AI for Sustainability in the Early Project Stages

Sustainability is an important issue to consider at the design stage, not only for environmental reasons, but also for economic and social reasons, as it promotes architectural quality and has economic benefits (Bragança et al. 2014). Construction activities still greatly affect environment, therefore developed countries committed to reducing the environmental impact of the construction sector (European Parliament 2024). Prioritizing sustainability in design aligns with the need for long-lasting solutions that promote well-being and minimize our dependence on natural resources like land, water, air, and energy.

Every decision made in the early stages of the design process serve as the bedrock for sustainable construction (Bragança et al. 2014; Robichaud and Anantatmula 2011). Early design choices establish the groundwork for subsequent design stages, these foundational decisions have a cascading effect, impacting factors like material selection, operational energy use, and construction waste generation. Later design modifications aimed at improving sustainability become increasingly challenging as the project progresses. It is safe to say that the most important steps towards achieving a sustainable building project within the established cost constraints are taken during the feasibility and design phases of the project (Robichaud and Anantatmula 2011). This is all the more true as the journey towards Near Zero Energy Buildings (NZEB) draws closer. Nowadays, the embodied energy overcomes the operational impact (Chastas et al. 2016; Grazieschi et al. 2020; Ramesh et al. 2010), so that the embodied carbon footprint is largely locked in by early design decisions, making significant reductions at a later stage very difficult or even impossible. Eventually, Integrating sustainability principles early in the design process fosters flexibility and improve overall system performance (Feria and Amado 2019). Deferring these considerations until later stages can restrict options and potentially necessitate expensive adaptations.

---

By Fulvio Re Cecconi.

© The Author(s), under exclusive license to Springer Nature Switzerland AG 2024
F. Re Cecconi et al., *Building Tomorrow: Unleashing the Potential of Artificial Intelligence in Construction*, PoliMI SpringerBriefs,
https://doi.org/10.1007/978-3-031-77197-2_3

The growing deployment of networked devices across urban environments is anticipated to result in a substantial increase in data generation. Consequently, big data analytics is increasingly recognized as offering exceptional approaches to addressing a range of emerging environmental and socio-economic challenges confronting modern cities (Bibri and Krogstie 2020). Data-driven approaches show greater efficiency and accuracy than knowledge-based approaches, which still have significant technical bottlenecks despite their interpretability and stability (Wan et al. 2023). (Cheng et al. 2020) highlight that data-driven techniques including machine-learning (ML) algorithms with big data are re-activating and re-empowering research in traditional disciplines for solving new problems. It's becoming increasingly important for all companies to be data-driven in order to develop effective tools for navigating the competitive terrain of their industries (Brynjolfsson et al. 2011), Márk and Ádám (2023), and the construction sector is no exception.

AI serves as an exceptionally effective tool for extracting valuable insights from the vast and complex datasets generated by the construction industry, which encompasses everything from project management and cost estimation to safety monitoring and equipment maintenance (Brynjolfsson et al. 2011). However, despite its potential benefits, the application of AI in this context is not without significant risks. One of the primary challenges lies in the quality of the data on which these AI models are trained. If the input data is incomplete, inaccurate, or biased, the AI systems can produce flawed outputs, leading to errors that may have severe and far-reaching consequences, such as financial losses, compromised safety, and project delays (Zammari and Ayob 2023). As such, the implementation of AI in the construction industry necessitates rigorous data governance practices to ensure that the data used is reliable, accurate, and representative of the diverse conditions and scenarios encountered in real-world construction projects.

In summary, it is clear that the recent technological innovations that the IT sector has brought to the construction world are providing the tools, and most importantly the data, to create new data-driven decision-making processes throughout the project lifecycle. For these new processes to be effective in improving sustainability in the sector, the decisions they enable need to be made at an early stage in the project. The remainder of this chapter discusses how big data and AI can be used to gain knowledge about the built environment, and how this knowledge can be used to make strategic decisions to decarbonise a building portfolio.

## 3.1  Leveraging Big Data to Understand the Built Environment

The design of buildings and, more generally, all decisions that have to be made to design, construct, use and manage the built environment are traditionally made on the basis of expert judgements that rely on their own knowledge, experience and specific studies generally based on a small data set (energy, structural, fire safety simulations, …). This causes a number of problems that traditionally plague the sector such as, among others, frequent cost overruns and delays in the design and construction of works (Hansen et al. 2023). This has become unacceptable in a sector that has a global market value of US$11.72 trillion in 2021 (Rafsanjani and Nabizadeh 2023) and constitutes one of the largest slices of every industrialised country's economy (Senaratne and Farhan 2023).

The digital revolution in the construction industry, known under the term Construction 4.0, has provided the technologies needed to collect and manage large amounts of data at all stages of a building's life cycle. The impact of the new generation of information technologies on the construction industry is not yet certain (Lee et al. 2021). However, these are developing rapidly and stimulate industrial transformation in an unprecedented situation (Xie et al. 2023). During the design and construction phases, digitisation has led to the spread of processes based on data-rich BIM models. In the remaining phases of the building and infrastructure life cycle, data collected with IoT tools have led to the spread of digital twins that generate gigabytes of data on a daily basis. In terms of applications, digital tools (BIM and DT) also find their way into a wide variety of sectors: buildings, railway infrastructure (Kaewunruen and Lian 2019) and bridges (Lu and Brilakis 2019). The availability of data makes the use of data-driven methods increasingly popular in various fields. A data-driven approach is one in which decisions are based on the analysis and interpretation of concrete data rather than on observation. A data-driven approach ensures that solutions and plans are supported by a range of factual information and not just intuitions, feelings and anecdotal evidence. Ease of use is probably the greatest of the many advantages of data-driven methods. Once data-driven models have been designed and tested, their application requires only the preparation of the necessary input parameters, without having to resort to physical reality modelling processes. These features make the data-driven approach particularly effective when dealing with physical models that are too complicated or require too expensive computing resources (Chen and Guan 2023). Furthermore, data-driven methods are preferred to traditional approaches because they promote transparency and accountability and continuously improve during use. Each new piece of data collected forms the pool of data on the basis of which decisions are made and thus reinforces the accuracy of the method.

For example, with the increasing availability and disclosure of building energy use data around the world, there is a growing opportunity to develop more effective and efficient data-driven models that can significantly assist urban planners and policy makers in making informed decisions related to energy management and sustainability (Jin et al. 2022). Such models not only have the potential to optimise energy use at the city scale, but can also be used to monitor critical issues such as energy poverty at the regional and local levels, allowing for targeted interventions and resource allocation to the most vulnerable communities (Terés-Zubiaga et al. 2023). However, the integration and application of these data-driven models is not without its challenges. A prominent concern among scholars is the uncertainty inherent in energy performance certificates (EPCs), which is often seen as a signifi-cant barrier to their adoption in building-level or spatial energy models (Hardy and Glew 2019). Despite these challenges, ongoing research efforts are aimed at reducing the uncertainty associated with EPC data to improve their reliability and accuracy. By refining the methods used to collect and analyse EPC data, researchers aim to make such data a more robust tool for determining the energy performance of build-ings, thereby improving the quality of inputs used in both urban energy planning and policy formulation (Wederhake et al. 2022). This growing body of work high-lights the potential of EPC data to contribute to more sustainable urban environments through better informed energy strategies.

As an example of how EPC data can be used to gain knowledge about residential buildings, an analysis of CENED DB (ARIA 2018) the energy cadastre of buildings in the Italian region of Lombardy, is presented. The open DB includes information on the primary and net energy performances of buildings, as well as geometric data (such as volume, gross and net surface, window area, etc.) and installed technologies (primarily data on the average thermal transmittance of building components and details on the overall efficiency of thermal plants). The dataset contains around 1.52 million records described by 45 features. Among these, the most relevant to this study are: (i) the gross and net heated surface,(ii) the gross and net volume; (iii) the envelope surface; (iv) the ratio between opaque and transparent envelope surface; (v) the average walls, windows, and roof thermal transmittance; and (vi) the primary energy for heating $EP_H$.

### 3.1.1  Data Cleaning

In Italy, the EPC of a building can be issued by a "certifying subject", who must be a qualified professional, usually architects, engineers and professionals without a university degree. The varying levels of education and skills among assessors lead to inconsistencies in EPC data, which undermines its reliability. As a result, this lack of standardization necessitates a rigorous data cleaning process to ensure that the data is accurate, comparable, and suitable for use in decision-making (Fig. 3.1).

The data cleaning process focused on single family houses. The DB contains more than 1.2 million records of residential buildings and more than 20% of them are clearly wrong. The main passages of the cleaning process had been:

1. Deleting all records not pertaining to residential assets
2. Deleting records with $EP_H \leq 0$ or $EP_H \geq 300$ kWh/m$^2$y
3. Deleting records with gross or net heated surface $\leq 20$ m$^2$
4. Deleting records with gross or net heated volume $\leq 30$ m$^3$
5. Deleting records with envelope surface $\leq 20$ m$^2$
6. Deleting records with the ratio between transparent and opaque envelope surface $\leq 0$

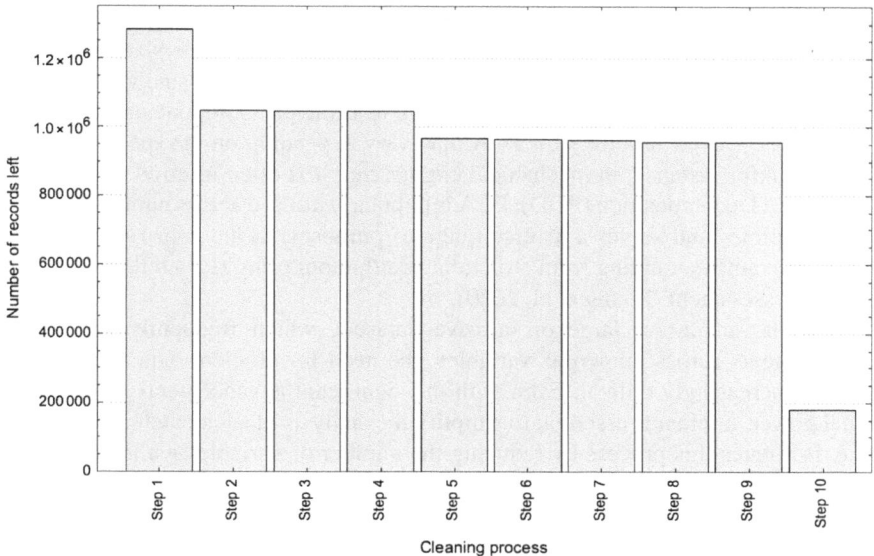

**Fig. 3.1**  Results of the data cleaning process on EPC data

7. Deleting records with average walls transmittance $U_{walls} \leq 0$ W/m$^2$K or $U_{walls} \geq 40$ W/m$^2$K
8. Deleting records with average windows transmittance $U_{win} <= 0$ W/m$^2$K or $U_{win} \geq 40$ W/m$^2$K
9. Deleting records with year of construction $y \leq 1800$ or $y \geq 2021$
10. Selecting single family houses

The data cleaning process then left 74% of the original dataset, of which approximately 19% were single family dwellings, and these were the focus of the research (Fig. 3.1). Two aspects of knowledge extraction from big data are illustrated here: (a) data dimensionality reduction using Principal Component Analysis PCA and (b) clustering building assets based on their PCA values.

## 3.1.2  Data Dimensionality Reduction

Data dimensionality reduction is a process in data analysis and machine learning that involves reducing the number of input variables or features in a data set while preserving as much relevant information as possible. This technique is critical for mitigating the curse of dimensionality (Köppen 2000), where high-dimensional data can lead to increased computational complexity, overfitting, and difficulties in visualising and interpreting the data.

Principal Component Analysis (PCA) is one of the most widely used dimensionality reduction methods. It is a statistical technique that employs an orthogonal transformation to convert a set of observations of potentially correlated variables into a set of linearly uncorrelated variables, known as Principal Components (Greenacre et al. 2022). The terminology for PCA may vary depending on the specific field of application, for instance, in mechanical engineering it is often referred to as Proper Orthogonal Decomposition (POD). PCA is popularly utilised across numerous scientific disciplines and serves a diverse range of purposes, in the construction sector there are examples ranging from structural health monitoring (Favarelli et al. 2022) to safety assessment (Zhang et al. 2020).

With the advent of large or massive datasets, which frequently encompass measurements across numerous variables, the need for efficient data analysis has become increasingly critical. Even with the significant advancements in computational power, it remains essential to simplify the analysis of such extensive datasets. PCA facilitates this process by reducing the number of variables while retaining a substantial amount of the information present in the original dataset. To achieve this reduction, PCA generates new variables, called Principal Components, which are derived as linear combinations of the original variables. The first Principal Component is determined to have the maximum possible variance, effectively capturing the greatest amount of information from the dataset. The second Principal Component is computed to be orthogonal to the first and is also designed to capture the maximum remaining variance. Subsequent components are calculated in a similar manner, each being orthogonal to all previous components and maximizing the variance within the constraints established by those components.

While PCA is a powerful tool for dimensionality reduction, it has limitations in cases where the data exhibits non-linear relationships. In many real-world applications the underlying data structure may be highly non-linear. In such instances, PCA may fail to capture the true data structure, resulting in suboptimal dimensionality reduction and poor performance in subsequent analysis or modelling tasks. To address these limitations, Kernel Principal Component Analysis (Kernel PCA) extends the conventional PCA by incorporating a non-linear transformation (Schölkopf et al. 1997). Kernel PCA utilizes the "kernel trick," which implicitly maps the original data into a higher-dimensional feature space where linear separability can be more easily achieved. By applying a kernel function, Kernel PCA can identify complex, non-linear patterns and project the data into a lower-dimensional space that better preserves the underlying structure. This makes Kernel PCA particularly useful in cases where the data is not linearly separable like in the case of EPC data. However, the use of Kernel PCA does present some challenges. One major difficulty is the selection of an appropriate kernel function and its associated hyperparameters, as different choices can significantly affect the results. In addition, Kernel PCA requires much more memory and computational resources than PCA, making it often impossible to run on standard computers, even for small to medium-sized datasets.

This is the case of the EPC dataset, which exceeded the resources available when attempting to reduce its dimensionality using Kernel PCA. It was therefore necessary to reduce the number of records in the dataset. This was done by randomly selecting

**Table 3.1** Comparison between the original dataset average of the features and the ones of the sampled dataset

| Feature | Original dataset | Sampled dataset |
|---|---|---|
| Gross heated surface (m$^2$) | 175.31 | 176.33 |
| Gross volume (m$^3$) | 573.19 | 577.45 |
| Gross envelope surface (m$^2$) | 379.83 | 381.40 |
| Ratio between windows and walls surface | 0.0688 | 0.0687 |
| Average walls transmittance (W/m$^2$K) | 0.8024 | 0.8034 |
| Average roofs transmittance (W/m$^2$K) | 0.7619 | 0.7624 |
| Average windows transmittance (W/m$^2$K) | 2.7536 | 2.7611 |
| CO$_2$ emissions (kg CO$_{2,eq}$) | 32.43 | 32.49 |
| Primary energy demand (kWh/m$^2$y) | 162.87 | 162.99 |

20% of the records, and all subsequent analyses were performed on this subset of the data. Table 3.1 shows that the sampling did not change the characteristics of the dataset, so the conclusions can be generalised to the whole building stock.

The PCA kernel revealed the presence of additional outliers (Fig. 3.2a) that had not been detected during the previous data cleaning. In the 2D histogram, almost all of the data is grouped into a very small rectangle (in yellow in the figure) in the lower left corner of the image. The histograms of the data projected onto the two principal components (on the right and top of the 2D histogram in Fig. 3.2a) clearly show how most of the data is grouped in a small region around the point (0, 0) in the two-dimensional space of the two principal components. A further step of data cleaning has been taken by dropping the data sets in the below 5% quantile and above 95% quantile on Principal Component 1 and the below 5% quantile and above 92.5% quantile on Principal Component 2. Figure 3.2b shows a 2D histogram of the cleaned data projected along the two principal components, highlighting with different colours the average primary energy demand (EP$_h$) of the records in each rectangle.

### 3.1.3  Clustering Dimensionally Reduced EPC Data to Gain Knowledge on the Building Portfolio

Clustering, an unsupervised machine learning algorithm (Celebi and Aydin 2016), is a standard approach for statistical data analysis used in many domains. It is the process of grouping a collection of objects so that items in the same group (called a cluster) are more similar than those in other groups (Xu and Wunsch 2005).

The clustering process is mainly characterised by two parameters: the clustering method and the number of clusters. Researchers typically follow a systematic approach, involving both theoretical considerations and empirical evaluation, to

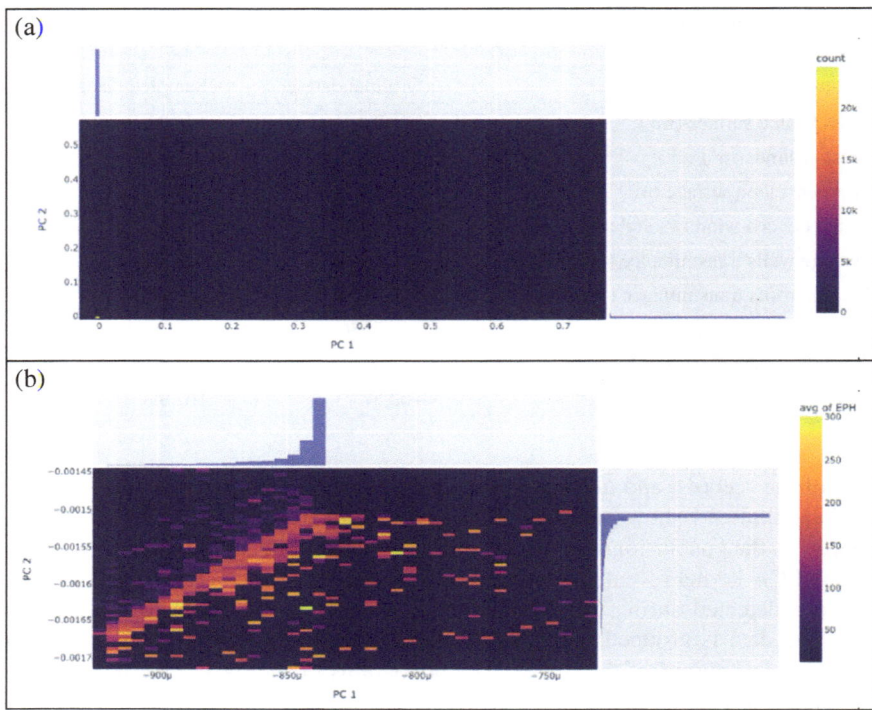

**Fig. 3.2** Two dimensions histogram of the dataset on the principal components. The picture at the top shows the number of assets before the last data cleaning step, the picture at the bottom the average primary energy demand after the data cleaning. The projections on the axes in both pictures show the number of assets

determine these parameters. The choice of clustering method depends on the nature of the data and the research objectives. Once a clustering method has been selected, the optimal number of clusters is determined using various validation techniques, such as the elbow method, silhouette analysis or gap statistics. The elbow method, used in this research, involves plotting the total intra-cluster sum of squares against the number of clusters and identifying the point where the rate of decrease changes sharply ('elbow') (Syakur et al. 2018).

For the EPC data, the K-Means algorithm was identified as the most effective clustering method. The optimal number of clusters, determined using the elbow method described earlier, was set to four. The key outcomes of the clustering process are illustrated in Fig. 3.3, where the average values of the primary characteristics for each cluster are displayed on a parallel axis graph.

One cluster, represented by the light blue line, consists of newer buildings (high Year of Construction), which are larger in both volume and surface area, with a substantial window surface (high windows-to-walls surface ratio) and very low average thermal transmittance for walls, roofs, and windows. These buildings exhibit

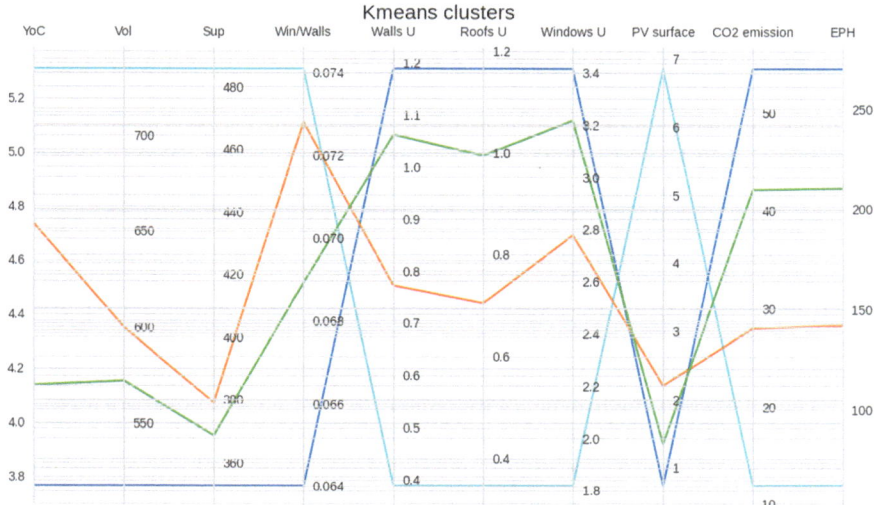

**Fig. 3.3** Parallel axis plot of the average of the main characteristics of the four clusters

the lowest energy demand and are characterized by the largest photovoltaic panel surface area. In contrast, the oldest buildings, indicated by the dark blue line, are the smallest in size and possess envelope components with the highest thermal transmittance. These buildings also have a very low average photovoltaic panel surface and, as expected, the highest energy demand. The remaining residential buildings are categorized into two clusters: one, represented by the green line, comprises older buildings with higher energy demand, while the other, denoted by the orange line, includes newer buildings with more efficient envelope components and lower energy demand.

Regardless of the specific outcomes, which would vary if the proposed methods of data cleaning, dimensionality reduction, and clustering were applied to a different building portfolio, it is important to emphasise that utilising existing data enables the efficient and cost-effective analysis of even very large building portfolios. This approach provides valuable insights at minimal cost, which would otherwise require significant investment in building surveys and performance measurements. The knowledge gained through this process facilitates informed political and strategic decision-making concerning the maintenance and refurbishment of the building stock. In the following chapter, it is demonstrated how a similar methodology, incorporating other machine learning tools, can lead to cost-effective, data-driven decisions for energy retrofitting.

## 3.2   ML for Strategic Decision Making on Energy Retrofitting Huge Building Stock

The uptake of energy retrofit investment in existing buildings has been limited, despite extensive empirical evidence of the environmental benefits of green buildings and the increasing urgency to reduce carbon emissions in cities (Lai et al. 2022). This urgency has been recognised by the European Union (EU), which has committed its member States to drastic reductions in pollution. This includes formulating and achieving new and consolidated strategies and directives for net-zero emissions by 2050 (European Commission 2018). To achieve this critical goal, targeted strategies have been implemented to reduce emissions in the most energy-intensive sectors, particularly in the buildings sector, one of Europe's largest energy consumers (Palladino and Turi 2023; United Nations Environment Programme 2022). Changes in energy use in the residential sector represent an essential segment of the ongoing low-carbon energy transition process, which is a multi-dimensional and multi-level process involving many actors (Ghafoori and Abdallah 2022).

A reliable tool to model and forecast building energy consumption is essential to address building energy efficiency issues and meet the current challenges of decarbonisation. It is of paramount importance for achieving smart and sustainable designs. Indeed, it can support more energy-efficient designs by comparing different strategies for both pre- (Celebi and Aydin 2016) and post-occupancy studies (Chastas et al. 2016). It can also guide energy management at local and global scales (Chen and Guan 2023). Among the three main approaches in building energy consumption modelling and forecasting (BECMF)—physics-based, data-driven, and hybrid models—data-driven techniques emerge as the most suitable option to ensure the integration of buildings in smart environments (Chen and Guan 2023; Cheng et al. 2020). Rather than relying on classical physics-based modelling tools, data-driven methods propose modelling and forecasting frameworks based on data analysis schemes. Furthermore, these frameworks include algorithms that benefit from the significant developments in machine learning in recent years, making modelling and predictive tools flexible and reliable. As a result, data-driven building energy performance modelling techniques have recently attracted increasing attention (Chen and Guan 2023; Dubey et al. 2019).

To speed up the identification of promising retrofit targets, this research proposes a data-driven methodology to forecast the economic feasibility of a building energy retrofit based on Energy Performance Certificates (EPC). EPCs are the most useful tools to evaluate the energy performance of buildings in standard conditions and to have energy information on national building stocks (Bibri and Krogstie 2020; ENEA 2023; European Commission 2018). Since EU Member States started collecting EPC data, the EPC databases have become one of the most important sources of information on building energy (ENEA 2023), although some concerns have been raised about the quality of the data (European Commission 2018).

Economic feasibility should be assessed before selecting a building retrofit alternative (Chastas et al. 2016). Identifying the optimal retrofit that achieves the best economic and environmental performance over the life cycle of a building is important due to financial constraints (European Parliament 2024). Nevertheless, evaluating retrofit projects based on initial cost alone has been recognised as a major drawback in asset management (Chastas et al. 2016; European Parliament 2024), as it does not consider the asset's operating costs, which can be significant over its lifetime. The Life Cycle Cost (LCC) analysis provides a method of determining the entire cost of an asset over its expected service life (SL) (Favarelli et al. 2022) along with operational and maintenance costs (Feria and Amado 2019). Since the first comprehensive introduction of LCC in the construction sector with a case study illustrating the payback of investments to reduce energy consumption (Gabrielli and Ruggeri 2019), there are many examples in the literature of its use in the evaluation of energy retrofit projects (Galimshina et al. 2020; Ghafoori and Abdallah 2022; Giuseppe et al. 2017; Goh et al. 2010; Grazieschi et al. 2020).

In summary, there is a need to improve the energy efficiency of the existing building stock through energy retrofits, the economic feasibility of which can be conveniently assessed using data-driven methods based on EPCs that measure costs over the life cycle of the assets. The article, after presenting the state of the art, introduces a data-driven method for early-stage decision-making in energy retrofit interventions. This method calculates the probability of achieving payback within 15, 20, or 25 years, providing a comprehensive tool for assessing the economic viability of retrofit projects in the context of long-term asset management.

### 3.2.1 Evaluating Cost-Effectiveness of Building Energy Retrofits: Savings Estimation and Comparative Methods

The cost-effectiveness of a building energy retrofit investment is calculated through economic evaluation methods based on LCC. These methods compare the costs of retrofitting with the savings over a defined period. Therefore, it is necessary to analyse the state of the art focusing on two different aspects: the estimation of the savings from the retrofit intervention, which depend directly on the lower energy consumption, and the methods for comparing them with the costs. These are the two topics for which the state of the art is examined in detail in this section.

*Predicting post-retrofit building energy performance*

Common approaches to predict building energy performance mainly include physical models ("white box"), hybrid methods ("grey box"), and data-driven approaches ("black box") (Greenacre et al. 2022). Physical models consider building materials' detailed physical properties and characteristics to predict thermal behaviour through numerical equations. Although physical models can describe heat transfer mechanisms, they suffer from the following disadvantages: (a) requiring expertise; (b) difficulty in making appropriate assumptions; (c) time consuming; and (d) inability to adapt to environmental/socio-economic vicissitudes. Hybrid methods combine physical models and data-driven approaches to simulate building energy (Hardy and Glew 2019), but they are subject to the same problems as physical models. These include inappropriate assumptions and the need for expertise (Greenacre et al. 2022). Data-driven approaches have shown the ability to overcome the above limitations of physical models and hybrid methods. This is due to their ability to discover statistical patterns from the available data set instead of physical information from the field (Greenacre et al. 2022). Therefore, in building energy prediction, data-driven approaches have recently attracted considerable attention (Cheng et al. 2020; Jin et al. 2022; Kingma and Ba 2017; Köppen 2000; Lai et al. 2022).

Gathering data and information is thus vital to leverage data-driven methods. Researchers think that the best sources for collecting the required information are generally the EPCs, which can be acquired from Government Energy Agencies and local public authorities (Márk and Ádám 2023). Although implemented differently in different countries in terms of the methodology of data collection and the features on which data are collected, EPCs generally include the following types of data features: (a) building reference (identification, type, construction year), (b) building geometry (floor area, envelope form), (c) certificate methodology (measured vs calculated data, time period of the audit), (d) factual energy consumption (energy use per source), (e) calculated energy performance, (f) energy system installations (HVAC, solar), (g) recommendations (and their implementation), (h) additional information (reference values, emissions) and (i) energy expert information (ENEA 2023).

A fairly recent study (ENEA 2023) reviews existing applications of EPC data and identifies thirteen application domains from a systematic mapping of 79 papers, revealing an increase in the number and complexity of studies and advances in applied data analysis techniques. One of these domains is the prediction of building energy demand. Although there is no consensus on the most appropriate tool for predicting energy consumption in buildings, in recent years various researchers have applied different modern tools, namely statistical or artificial intelligence (AI) tools, which are gaining momentum thanks to advances in the AI field (Moolayil 2019; Palladino and Turi 2023; Pernetti et al. 2021; Ramesh et al. 2010; Robichaud and Anantatmula 2011; Schölkopf et al. 1997). Of all the techniques offered by AI, Artificial Neural Networks (ANN) are the most widely used for predicting building energy consumption (Moolayil 2019) due to the exceptional results they may produce when trained on a large amount of data. ANN is a non-linear computational model that emulates the functional concepts of the human brain (Syakur et al. 2018). The basic form of ANN consists of three successive layers, namely input, hidden and output layers (shallow neural network). The input layer is used to feed the model, the hidden layer is the bridge between the input and output layers and contains most of the parameters to be trained, while the output layer provides the result. The more complex the problem, the more neurons should be added to the hidden layer. The evolution of shallow networks is the deep neural network, where the number of hidden layers increases. Section 3.3 illustrates the use of an ANN to predict the energy consumption of a building using data from the EPCs of a building portfolio. ANNs have also been used in this study to predict the economic feasibility of retrofitting a building, in particular, to calculate the reliability of a given payback period (Sect. 3.2.5).

*Life cycle costing*

Life cycle costing is concerned with optimising value for money in the ownership of physical assets by considering all the cost factors relating to the asset during its service life (Terés-Zubiaga et al. 2023). It was first developed in the mid-1960s to assist the US Department of Defence procure military equipment (United Nations Environment Programme 2022). From the mid-1980s, most US government agencies were required to use formal life-cycle evaluation methods, and many private owners chose to use these methods in making building investments in order to assess and compare the relative benefits of alternative energy design options in buildings (Wan et al. 2023). From the standards point of view, the LCC also has a long history, and the two most important references, the American one, ASTM E 917 (Wederhake et al. 2022), and the international one, ISO 15686–5 (Wu et al. 2016), are worth mentioning here. The former is a practice that establishes a procedure for evaluating the life-cycle cost (LCC) of a building or building system and comparing the LCCs of alternative building designs or systems that satisfy the same functional requirements (Wederhake et al. 2022).

LCC analysis covers a defined list of costs over the physical, technical, economic or functional life of a constructed asset over a defined analysis period (Wu et al. 2016). All relevant costs associated with owning and operating a building or building system are measured in present value, i.e., the value found by discounting future cash flows to the base time (Xu and Wunsch 2005), or annual value, i.e., uniform annual amount equivalent to the project costs or benefits taking into account the time value of money throughout the study period (Xu and Wunsch 2005), terms. Several studies demonstrated that estimating the life cycle economic performance by summing all the impacts incurred during each life cycle stage over a lifetime is the most straightforward and commonly employed method for comparing building performances.

Besides LCC, there are many other methods to assess the economic performance of a project. Most are based on the same inputs as LCC and form a family of economic evaluation methods, including benefit–cost and savings-investment ratios, net benefits, internal rates of return, and payback periods. The Payback Period (PBP) is, according to (Xu and Wunsch 2005), the time required for the cumulative benefits from an investment to pay back the investment cost and other accrued costs considering the time value of money. Despite its inherent limitations, the payback period is a widely used metric for assessing the economic performance of energy retrofit projects.

The PB may be computed by solving Eq. (3.1):

$$\sum_{t=1}^{PB} \frac{\left( B_t - \tilde{C}_t \right)}{(1 + i)^t} = C_0 \tag{3.1}$$

where:

$(B_t - \tilde{C}_t)$ is the net cash flow in year t computed as the difference between the value of benefits in year t $B_t$ minus the value of costs in year t $\tilde{C}_t$.

$i$ is the discount rate per time period.

$C_0$ are the initial project investment costs.

Noteworthy, Eq. (3.1) may not have a solution. This is the case when the savings. i.e., the left part of Eq. (3.1), over the study period, are insufficient to pay off the initial investment.

Life cycle costing and all the associated methods like PBP are a prediction of the future, and therefore, different cost estimation methods need to be used. Different cost estimation methods depend, for example, on the availability of data and the stage at which the calculations are made. There are at least three different ways to estimate costs: (a) estimating by engineering procedures; (b) estimating by analogy; and (c) parametric estimating methods. Regardless of the method, due to buildings' long service lives, many uncertainties might cause a deviation in the results of a predicted retrofit outcome. Many sources of uncertainty can be identified in the analyses. These include investment costs, energy prices, component replacement times, inflation rates etc. The combined effect of these uncertainties can lead to a significant deviation from the expected result of the LCC analysis. Many techniques exist, and several studies using different methodologies have been carried out to model the uncertainties and quantify their impact in the LCC analysis of energy retrofit projects. Monte Carlo Simulation (MCS) is the most commonly used method to deal with uncertainties in the input values of LCC analyses. MC simulation uses computational power to explore all possible outcomes of a problem, given certain bounds on the variability expressed in the model. Its main advantage over deterministic models is that it allows uncertainty and risk during the long-term operation of buildings to be incorporated into cost analyses. This technique can be used in life cycle modelling to discover the likely variability in the expected outcome. Section 3.5 details how MC simulations were used in this research, showing how the probability distributions of the input variables were chosen and which literature references were consulted for each input.

### 3.2.2  Predicting the Economic Feasibility of an Energy Retrofit

The methodology for predicting the financial feasibility of energy retrofit interventions meets three key criteria: (a) it is designed for use in early decision-making stages, where its impact is most significant; (b) it caters to users not specialized in energy modelling; and (c) it produces results easily interpretable by investors. To fulfil these needs, payback time was chosen as the metric for economic feasibility, considering both the investment cost and the time value of money. A black-box approach, specifically using artificial neural networks, was opted for due to its simplicity and minimal data requirement, making it accessible to a wider audience. This approach led to the development of a tool based on Energy Performance Certificates (EPC) data, capable of predicting the payback period of 15, 20 or 25 years for an energy retrofit investment with limited input. Figure 3.4 shows the main steps to obtain the forecasting tool.

The overall schema can be summarised in the following steps:

- creation of a tool capable of predicting the energy demand of a building after retrofit. This tool is necessary to know the potential savings in heating costs. This tool, detailed in Sect. 3.2.3, utilizes EPC data, offering broader applicability than models based on physical building behaviour. Which are generally more complex and require more inputs and computing time
- Development of a database detailing energy retrofit interventions and their maximum viable investments for payback periods of 15, 20, or 25 years. This database, described in Sect. 3.2.4, used the predictive tool to generate a list of retrofit scenarios for the assets included in the CENED DB. This step is essential for creating a comprehensive dataset for the next phase.
- Construction of a tool capable of predicting the likelihood of achieving desired payback periods (15, 20, or 25 years) after an energy retrofit. This tool, elaborated in Sect. 3.2.5, utilizes three neural networks to predict the probability of achieving the desired payback period and is based on the dataset from the preceding steps.

These three main steps will be discussed in the remainder of this chapter, with the final part devoted to discussing the results obtained and drawing some conclusions.

### 3.2.3  Machine Learning for Predicting Post-retrofit Energy Savings at Portfolio Scale

The primary objective of this research, depicted in Fig. 3.3, was to develop a tool for estimating the energy demand of various buildings using data from Energy Performance Certificates (EPCs) in Regione Lombardia, sourced from the CENED DB (ARIA 2018). The analysis focused on residential buildings, encompassing a variety of property types including apartments, single-family homes, and villas, enabling broad applicability of the tool for assessing energy performance.

A key challenge was addressing data quality issues, including inconsistencies and outliers in the distribution of primary energy demand by energy label and construction year, as shown in Fig. 3.5. These discrepancies arose from variations in the expertise of certificate issuers and inadequate data control. Effective data cleaning was essential to ensure the reliability of the analysis, as undetected errors can lead to misleading results and decisions.

This process started with selecting records pertaining only to residential properties from the 1.52 million records in the CENED DB (Eq. 3.2):

$$\text{SELECT} * \text{WHERE N} = \text{"E.1(1)" OR N} = \text{"E.1(2)"} \tag{3.2}$$

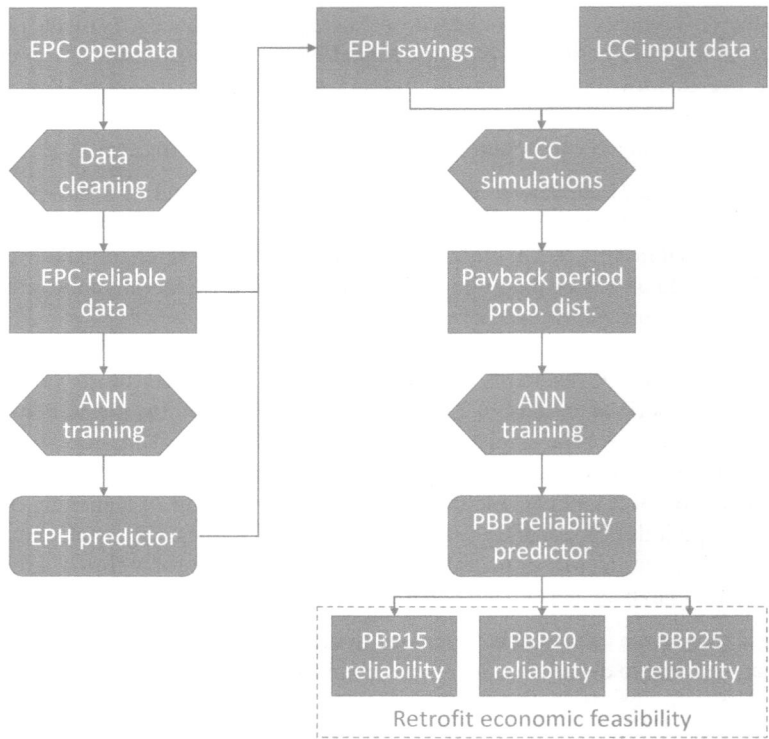

**Fig. 3.4**  Research schema

The two strings for column "N", classified according to the Italian Law DPR 412/ 1993, allow to select assets with two types of intended use embracing: (a) "E.1 (1)" dwellings used as a continuous residence, such as civil and rural dwellings, colleges, convents, penal institutions, and military barracks; (b) "E.1 (2)" dwellings used as residences with occasional occupancies, such as holiday homes, weekend homes and the like.

The analysis was then narrowed to single-unit buildings, excluding flats in multi-unit buildings, to maintain consistency in the types of energy retrofits and associated costs studied. The research scope was significantly impacted by this decision, as energy retrofit possibilities and costs vary greatly between single-family and multi-family dwellings. The database was reduced by almost 25% during the subsequent data cleaning process, which followed the methodology described in Re Cecconi et al. (2022). As a result, just over 161,000 records of single-family dwellings were included in the study. These records were then used to train an Artificial Neural Network (ANN) to predict a building's primary energy demand.

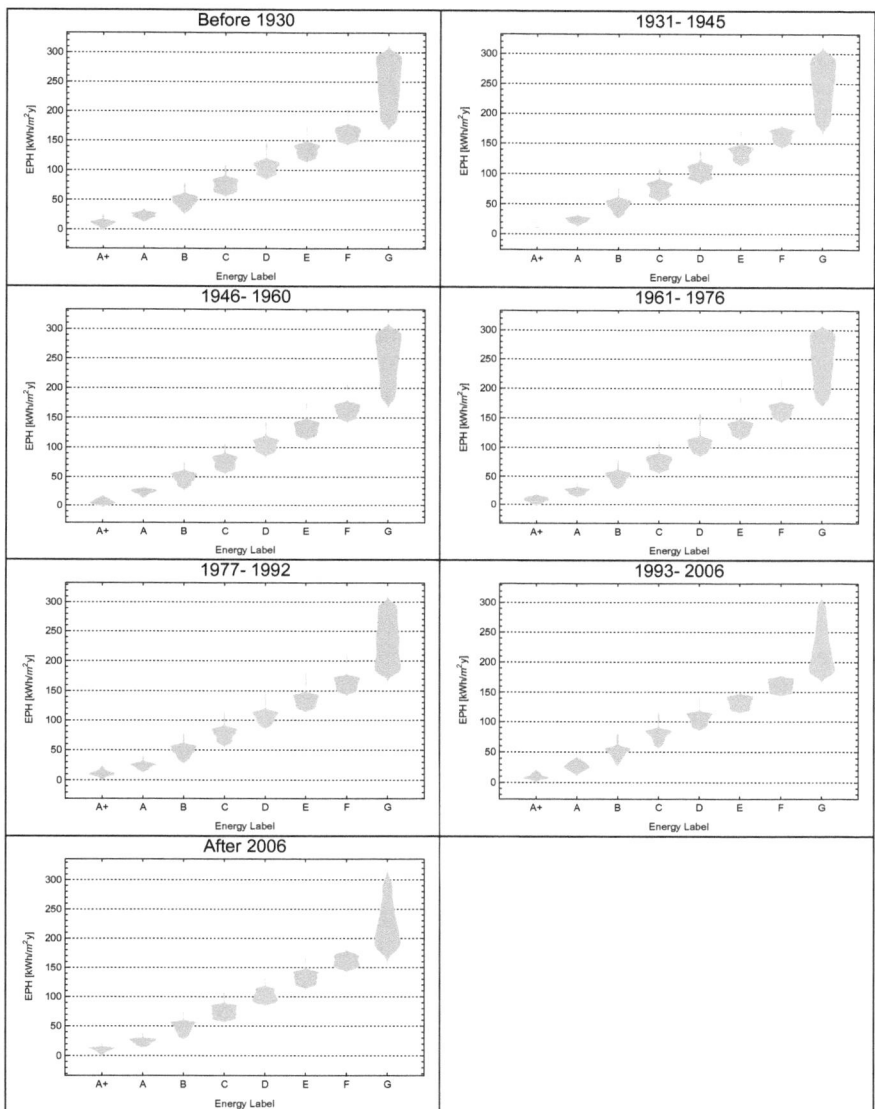

**Fig. 3.5** Distribution of the $EP_h$ by year of construction of the asset and its energy label

*ANN training*

The first step in implementing an ANN model is the selection of meaningful features. These features must form a logical input set for the model; therefore, having knowledge of the domain is fundamental at this stage. The CENED DB has 44 features, including the proposed ANN output, $EP_h$. Many of them are not necessary to reliably predict the primary energy demand of the building. Describing the feature selection process is beyond the scope of this article; here are the selected ones: (a) City name; (b) Year of construction; (c) Gross heated volume; (d) Dispersion surface; (e) Glass to wall surface ratio; (f) Wall average transmittance; (g) Roof average transmittance; (h) Window average transmittance. The first two are categorical variables, while the others are numerical variables whose statistical description is given in Table 3.2.

Before the training phase, other transformations on the dataset were necessary: for instance, the "YEAR_OF_CONSTRUCTION" column was transformed into a numerical value according to seven categories: "Before 1930", "1931–45", "1946–60", "1961–76", "1977–92", "1993–2006", and "After 2006" (as in Fig. 3.2). The dataset was then split into three subsets: training, validation, and test, using the following proportion of the original dataset: 70, 10, 20% respectively. The training and validation subsets are used to monitor the model's performance during training and prevent overfitting. The test set is used to test the performance of the model in unseen data.

Developing an energy consumption predictor using ANNs involves several stages:

- Network Architecture Design: Defining the neural network structure, including the number of layers and neurons, and the activation functions. The specific architecture used in this research, comprises six layers: a normalisation layer, four hidden layers up to 256 networks, and one output layer.
- Training and Optimization: The network is trained on the pre-processed data, iteratively adjusting its weights to minimize prediction error. The Adam gradient-based optimization algorithm was used for this purpose (Kingma and Ba 2017).
- Hyperparameter Tuning: involves the systematic optimization of parameters that control the learning process but are not updated during training. These hyperparameters include the learning rate, batch size, and activation functions, among others.

Finally, the model's performance was assessed using the Mean Absolute Error (MAE) metric. The final model achieved an MAE of 17.78, indicating an error smaller than the 5% quantile of the training dataset. Figure 3.6 displays the comparison between predicted and actual $EP_h$ values, with a diagonal trend indicating the model's ability to accurately infer $EP_h$ distributions based on the selected parameters.

## 3.2.4  Energy Retrofit Life Cycle Costing

In the second step of this research, potential energy retrofit scenarios for each building in the CENED database are identified, and the maximum recoverable investment cost over 15, 20, or 25 years is calculated. This process yields a comprehensive database of energy retrofit measures along with their corresponding payback periods, which is essential for the final phase of the study. For each retrofit scenario, the energy savings are calculated using the neural network developed in the first step. A key aspect of this phase is addressing the uncertainties in economic and financial parameters

**Table 3.2** The statistical description of the numerical features of the dataset used to train the ANN

| | | Gross volume (m³) | Dispersing surface (m²) | Glass over walls surface ratio | Walls average transmittance (W/ m² K) | Roofs average transmittance (W/ m² K) | Windows average transmittance (W/ m² K) | Eph (kWh/ m² y) |
|---|---|---|---|---|---|---|---|---|
| Mean | | 708.71 | 413.86 | 0.073 | 0.873 | 0.840 | 3.043 | 168.24 |
| Standard deviation | | 1'090.00 | 485.00 | 0.034 | 0.455 | 0.508 | 0.989 | 78.36 |
| Min | | 58.60 | 21.38 | 0.001 | 0.010 | 0.001 | 0.629 | 0.02 |
| Quantile | 5% | 174.90 | 106.04 | 0.035 | 0.244 | 0.205 | 1.385 | 40.69 |
| | 25% | 292.68 | 187.16 | 0.052 | 0.503 | 0.389 | 2.302 | 106.01 |
| | 50% | 424.00 | 280.51 | 0.066 | 0.810 | 0.700 | 3.230 | 169.46 |
| | 75% | 642.91 | 438.66 | 0.085 | 1.197 | 1.300 | 3.382 | 235.37 |
| | 95% | 2'284.02 | 1'213.87 | 0.131 | 1.634 | 1.700 | 4.920 | 287.35 |
| Max | | 9'999.00 | 11'330.62 | 0.540 | 2.604 | 6.897 | 6.478 | 300.00 |

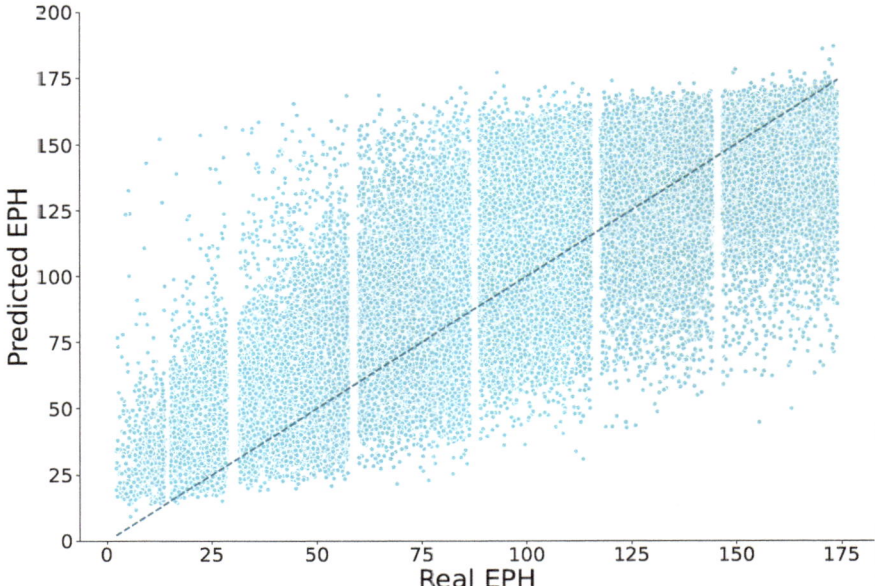

**Fig. 3.6** Predicted EPH vs Real EPH. Noteworthy, the real EPH values present vertical lines with no records near the limit values of each energy label. This trend results from the manual process carried out by the experts, who prefer to report values so that they are easily ascribed to a specific category and are not on the borderline between two different categories

critical to retrofit cost calculations. Several studies (Baldoni et al. 2021; Gabrielli and Ruggeri 2019; Galimshina et al. 2020; Giuseppe et al. 2017; Goh et al. 2010) have highlighted the risks of errors in life cycle cost calculations due to underestimated uncertainties. The same studies provide appropriate statistical distributions that can be used to model the uncertainty associated with each input variable in the LCC analysis. Table 3.3 summarises the assumptions made to model the uncertainty for all the inputs in the LCC analysis, with the chosen statistical distributions mainly taken from Galimshina et al. (2020).

To account for uncertainties in Energy Performance ($EP_h$) predictions, a Gaussian distribution was utilized, centred around the $EP_h$ value predicted by the ANN, with a standard deviation associated with the Mean Absolute Error (MAE) from the training phase, as indicated in Eq. (3.3).

$$\sigma = MAE \times \sqrt{\frac{2}{\pi}} \tag{3.3}$$

The distribution is truncated to avoid negative $EP_h$ values, which are physically implausible, covering more than 99.7% of potential outcomes within a range of six standard deviations.

**Table 3.3** Economic and financial inputs for life cycle costing

| Quantity | Probability distribution function | Probability distribution function parameters |
|---|---|---|
| Primal energy demand $E_{ph}$ | Truncated Gaussian | Mean $= E_{ph,predicted}$ |
| | | Standard deviation $[\sigma_{Eph}] = 19.775$ |
| | | Min $= \{0$ $if$ $E_{ph,predicted} - 3 * \sigma_{Eph} <$ $0$ $E_{ph,predicted} - 3 * \sigma_{Eph}$ $otherwise$ |
| | | Max $= \{2 * E_{ph,predicted}$ $if$ $E_{ph,predicted} - 3 * \sigma_{Eph} <$ $0$ $E_{ph,predicted} + 3 * \sigma_{Eph}$ $otherwise$ |
| Energy cost | Weibull | Shape $= 1.5$ |
| | | Scale $= 0.12$ |
| Nominal discount rate | Triangular | Min $= 1\%$ |
| | | Peak $= 4\%$ |
| | | Max $= 7\%$ |
| Inflation | Truncated Gaussian | Mean $= 3\%$ |
| | | Standard deviation $[\sigma_{infl}] = 0.01$ |
| | | Min $= 0$ |
| | | Max $= 6 * \sigma_{infl}$ |

The same type of statistical distribution was used to model inflation, Table 3.3 shows the numerical values used for mean, standard deviation, minimum and maximum. Energy costs are modelled using a Weibull distribution with shape 1.5 and scale 0.12. Finally, the nominal discount rate is modelled using a triangular distribution with a minimum of 1%, a maximum of 7% and a maximum of 4%.

It's important to note that these specific assumptions about input distributions for calculating life cycle savings and costs, and thus the energy retrofit's payback period, are tailored to the current economic and market conditions in Italy. However, the proposed methodology is versatile and can be adapted to other countries by altering the input parameters, demonstrating its broad applicability.

*Computing retrofit payback probability distributions*

In calculating the payback period (PB) using Eq. 3.1, several simplifying assumptions were made. First, annual maintenance and repair costs post-retrofit were assumed to be comparable to those pre-retrofit. This assumption is justified by the balancing effect of increased maintenance costs for new systems and reduced maintenance costs for the renovated building components, as well as the costs associated with repairing outdated and damaged systems. Second, replacement costs were presumed to remain unchanged before and after retrofitting. Lastly, the residual value of these investments was considered to be zero.

With these simplifications, the right side of Eq. 3.1 represents the retrofitting investment cost, which serves as an input to the research tool discussed in Sect. 3.2.3 and is compared to the discounted cash flow (represented by the left side of Eq. 3.1), incorporating the uncertainties previously outlined. To account for these uncertainties in calculating the cash flow, Monte Carlo (MC) simulations were employed, as indicated in the existing literature. MC simulations generate random data sets based on probability distributions (outlined in Table 3.3) and analyse these sets to determine outcomes such as the discounted cash flow for each payback period, thereby facilitating the understanding of complex systems and informed decision-making under uncertainty.

A single MC simulation with 5,000 iterations was conducted for each retrofit scenario of every asset in the CENED database. These scenarios ranged from a comprehensive retrofit to achieve an A+ energy class to a minimal retrofit focused primarily on insulation, along with five intermediate options. The outcome comprises millions of cash flow values, which, when collectively analysed, illustrate the probability distribution of the discounted cash flows for each asset, corresponding to the left term of Eq. 3.1. For example, Table 3.4 illustrates the results of simulations using a 15-year PBP. There are 7 violin plots in the figure on top of the table, one for each previously identified "year of construction" class. They show the 75% quantiles of the discounted cash flow distributions obtained with the MC simulations. Notably, buildings constructed after 2006 show significantly lower distributions, indicating lower average savings from retrofits compared to older buildings. Older 'YEAR_OF_CONSTRUCTION' classes, on the other hand, have more similar distributions. The table also shows the statistical description of the 25, 50, 75 and 95% quantiles made by the principal moments of the distribution. Same information for 20 and 25 years of payback time were gained. These probability distributions help in predicting the reliability of energy retrofit payback times for different assets.

### 3.2.5   Predicting Energy Retrofit Payback Time Reliability Using Artificial Neural Networks

The previous steps provided a dataset of possible energy retrofits for a portfolio of assets with seven types of retrofit strategies for each building. A probability distribution of the maximum investment for a payback period of 15, 20, or 25 years was calculated for each type of retrofit.

This process mirrors what engineers typically encounter when evaluating the cost-effectiveness of energy retrofits in buildings. However, replicating this in current practice is challenging, requiring extensive knowledge of building energy behaviour and proficiency in techniques like Monte Carlo simulations to address uncertainties in input factors. Moreover, this traditional approach is often time-consuming and costly, and not feasible in the early project stages (project initiation, feasibility study, project definition and concept design) due to the requirement of detailed inputs.

The intricacies of energy models, demanding specialised knowledge, and the mathematical skills needed to handle the uncertainty in the LCC input can be circumvented by leveraging a black-box model such as neural networks. A trained ANN

**Table 3.4** Statistical description of the quantiles of the savings (Euro/m$^2$) given a payback period of 15 years. The image shows the distribution of quantile 75% values according to the year of construction of the building. The short horizontal lines of each violin diagram show: (a) the maximum and minimum; (b) the mean and median; (c) the quantiles 5, 25, 75, 95%

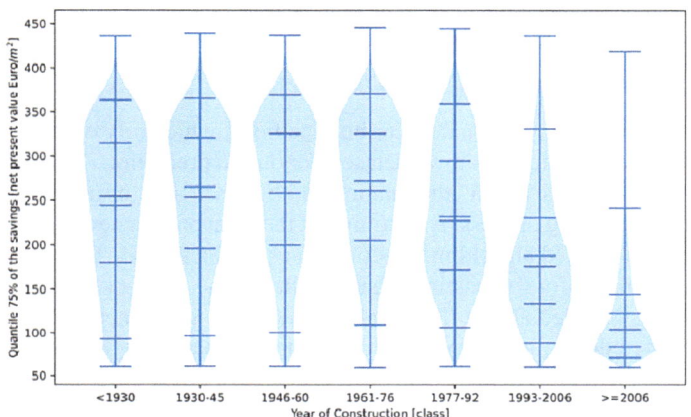

| PBP15 | Q 25% | Q 50% | Q 75% | Q 95% |
|-------|-------|-------|-------|-------|
| Mean  | 67.12 | 121.43 | 195.19 | 335.52 |
| Std   | 37.81 | 67.12 | 105.79 | 176.93 |
| Min   | 0.00  | 0.00  | 0.00  | 0.00  |
| 25%   | 37.03 | 67.41 | 109.20 | 190.78 |
| 50%   | 68.09 | 122.99 | 197.14 | 337.18 |
| 75%   | 98.46 | 177.67 | 284.15 | 483.96 |
| Max   | 170.95 | 279.47 | 445.72 | 772.23 |

can be used even by those without specialized knowledge, simplifying the process significantly. We found the optimal balance between the complexity of input data and the reliability of results by estimating four threshold values for the maximum allowable investment for the payback period. These thresholds correspond to the 25, 50, 75, and 95% quantiles of the probability distributions from the Monte Carlo simulations. The ANN uses the following inputs to achieve this result:

- Year of Construction (in 8 classes as in CENED DB);
- Name of the city where the building is located (automatically transformed in heating degree days);
- Building gross volume (m$^3$);
- External surface of the building (m$^2$);
- Ratio between the surface of windows and the surface of external walls;
- Actual average thermal transmittance of walls, roofs, and windows (kW/m$^2$K);

– Target (post-retrofit) average thermal transmittance of walls, roofs, and windows ($kW/m^2K$).

These input parameters do not require much knowledge of the building's energy behaviour. Moreover, they can be estimated in the first stages of the decision-making process. This means that decisions about energy retrofits can be made early, maximizing their potential impact in terms of energy savings.

*Training the energy retrofit payback time reliability predictor*

Neural networks differ from conventional white-box methods, which rely on prede-fined mathematical models applied to known datasets for output prediction (Wu et al. 2016). Instead, Neural networks "learn" the relationship between a set of input data and output data during a training process. After training, they can predict new outputs from new data without a defined algorithm, functioning as a 'black-box'. The selection of neural network inputs for forecasting the reliability of the payback period of a retrofit intervention is detailed above. The final design of the neural networks, as shown in Table 3.5, includes an input layer with 11 neurons (matching the number of inputs), seven hidden layers with varying neuron counts, and an output layer with four neurons. These four neurons correspond to the 25, 50, 75, and 95% quantiles of the maximum allowable investment payback periods derived from Monte Carlo simulations. The Leaky Rectified Linear Unit (LeakyReLU) activation function is employed in all layers of the model. It is a variation of the Rectified Linear Unit (ReLU) and is designed to allow a small gradient even when the unit is inactive. This feature helps to address the vanishing gradient problem, which can otherwise impede the training process (Dubey et al. 2019; Moolayil 2019).

Three identical neural networks were trained for the payback periods of 15, 20, and 25 years. Figure 3.7 shows the Mean Absolute Error (MAE) on both the training and validation datasets, indicating that the networks are not overfitted—they have learned to generalize to new data effectively.

**Table 3.5** Structure of the three ANNs to predict the probability of the payback time. Each net has 547.748 trainable parameters

| Layer | Number of neurons | Number of parameters |
|---|---|---|
| Input (normalization) layer | 11 | 23 |
| Hidden layer | 32 | 384 |
| Hidden layer | 64 | 2.112 |
| Hidden layer | 128 | 8.320 |
| Hidden layer | 2048 | 264.192 |
| Hidden layer | 128 | 262.272 |
| Hidden layer | 64 | 8.256 |
| Hidden layer | 32 | 2.080 |
| Output layer | 4 | 132 |

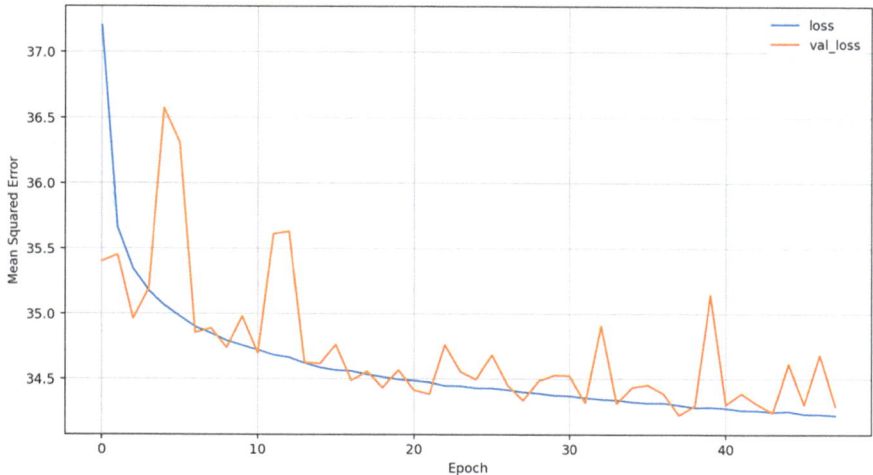

**Fig. 3.7**  Training loss (blue line) and validation loss (amber line) measured with the Mean Absolute Error (MAE)

Figure 3.8 provides an additional assessment of the trained networks' capabilities for a payback period of 15 years. Here, four scatter plots display the comparison between test values (Truth) and network-calculated results (Predicted) for each of the quantiles that the network was designed to predict. These scatter plots exhibit high correlation between the network predictions and the actual test values, as seen from the dense diagonal plots. It should be noted that the test results have never been seen by the network during the training phase and are therefore entirely new to the net. Similar results have been obtained for the other two payback periods.

The networks' training outcomes are summarized in Table 3.6, which presents Pearson's correlation coefficients for the test and predicted data.

Despite a limited number of inputs, the networks show a remarkable correlation with actual values, with coefficients above 0.81. This high correlation is significant, especially considering the complexity of accurately predicting investment return times. Additionally, Fig. 3.8 indicates that the networks tend to make conservative predictions, usually forecasting slightly lower maximum investment values than actual, which is advantageous for providing users with cautious estimates.

*Using the energy retrofit payback time reliability predictor*

The three trained neural networks form the backbone of the tool for assessing the reliability of predicting the payback period of an energy retrofit investment in a single-family residential building. The user, i.e. the investor, must feed the networks with information about the building:

- Year of Construction (in 8 classes as in CENED DB);
- Name of the city where the building is located (automatically transformed in heating degree days);

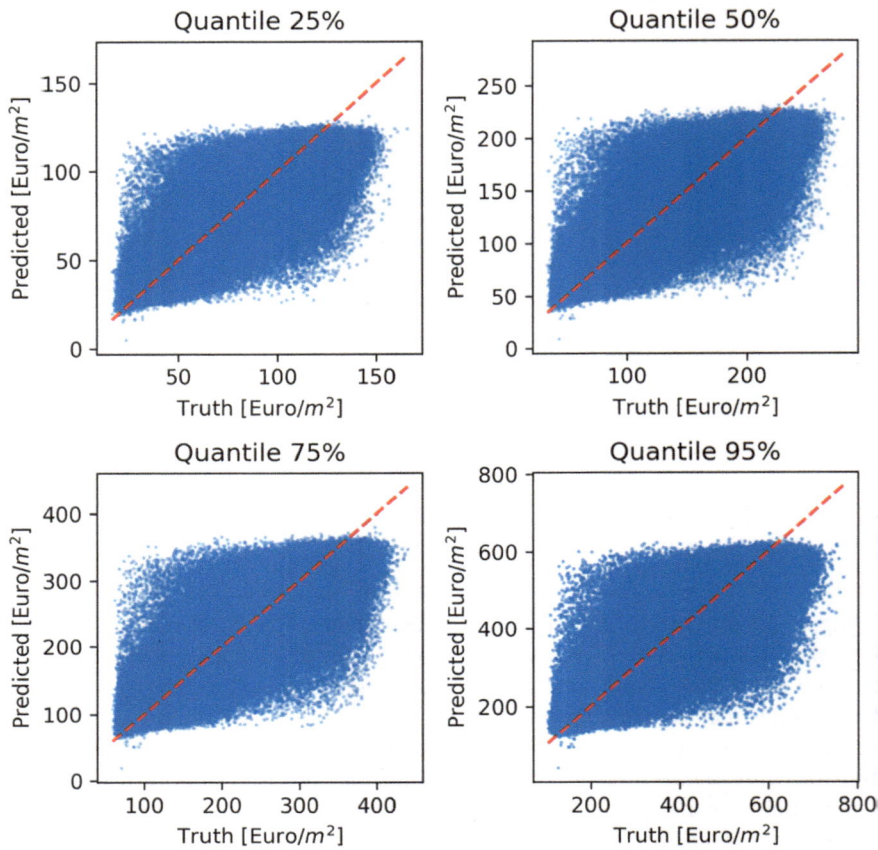

**Fig. 3.8** ANN performances on test data for PBP15

**Table 3.6** Pearson's coefficient measuring the correlation between the predicted and the true value on the test dataset

|        | Q25%       | Q50%        | Q75%       | Q95%       |
|--------|------------|-------------|------------|------------|
| PBP15  | 0.81800529 | 0.822309568 | 0.82426697 | 0.82176473 |
| PBP20  | 0.81803817 | 0.82333668  | 0.82479039 | 0.82261771 |
| PBP25  | 0.81871132 | 0.8240822   | 0.82524313 | 0.82240961 |

– Building gross volume (m$^3$);
– External surface of the building (m$^2$);
– Ratio between the surface of windows and the surface of external walls;

The investor must also enter information summarising the planned energy retrofit:

– Actual average thermal transmittance of walls, roofs, and windows (kW/m$^2$K);

– Target (post-retrofit) average thermal transmittance of walls, roofs, and windows (kW/m$^2$K).

Given this input data, the networks provide the four quantiles of the maximum investment that can be paid off in the three payback periods. To make the output of the forecasting tool easier to be understood, it is presented in a diagram as shown in Fig. 3.9. It shows three columns, one for each payback period, on which the investment thresholds corresponding to the four quantiles calculated by the network are represented in different colours.

Investors can use this diagram to determine the likelihood of achieving their desired payback period. By selecting a payback period and expected retrofit expenditure, the diagram indicates the probability of meeting the payback period. Different colour zones represent varying probabilities: green for a probability greater than or equal to 75%, light green for 50–75%, light yellow for 25–50%, amber for 5–25%, and red for less than 5%. Beyond individual buildings, this tool can efficiently process data for entire building portfolios. For instance, using the CENED database, it quickly calculates the maximum investment for converting all single-family residential buildings into Nearly Zero Energy Buildings (NZEB) within specific payback periods.

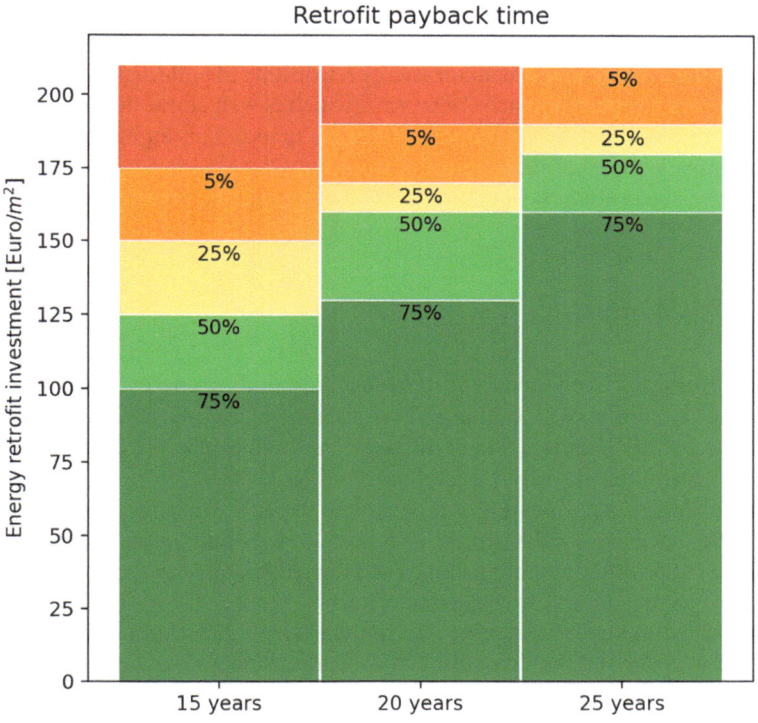

**Fig. 3.9** Output of the retrofit payback time reliability predictor

**Table 3.7** Maximum retrofit investment (Euro/m$^2$) that can be paid off in 15, 20 and 25 with a 75% reliability: mean value computed on all the single family buildings in the CENED DB

|        | <1930 (Euro/ m$^2$) | 1930–45 (Euro/m$^2$) | 1946–60 (Euro/m$^2$) | 1961–76 (Euro/m$^2$) | 1977–92 (Euro/m$^2$) | 1993–2006 (Euro/m$^2$) | >2006 (Euro/ m$^2$) |
|--------|-------|---------|---------|---------|---------|-----------|--------|
| PBP15  | 70.83 | 74.74   | 76.26   | 78.58   | 70.03   | 55.22     | 14.93  |
| PBP20  | 82.14 | 86.82   | 88.31   | 91.15   | 81.24   | 64.06     | 17.32  |
| PBP25  | 90.05 | 95.23   | 97      | 99.99   | 89.14   | 70.25     | 19.02  |

Table 3.7 shows the average investment value with a 75% probability of payback, highlighting that even for older buildings, the maximum investment is relatively low, typically under 100 Euro/m$^2$.

This value can easily be compared to the amount spent on the energy retrofitting of single-family buildings in Lombardy over the past two years thanks to a tax incentive campaign promoted by the Italian government. According to a recent report (ENEA—National agency for new technologies 2023) published by the National Agency for New Technologies, Energy and Sustainable Economic Development (ENEA), more than 33,000 single-family buildings in the Lombardy have undergone energy retrofitting thanks to tax incentives with an average investment of 120,946.95 Euros. The average surface area of residential buildings in the Lombardy computed from data in the CENED DB is about 177 m$^2$. The combination of these two data shows that the average investment for energy retrofitting of buildings in the Lombardy in recent years has been about 702 Euro/m$^2$, which is well above the average limit of 100 Euro/m$^2$ calculated by the proposed forecasting tool. This highlights that without such incentives, many of these retrofits would not have been economically viable, underscoring the usefulness of our tool in providing realistic investment assessments.

### 3.2.6  Discussions

In line with Europe's commitment to enhancing building energy performance and reducing $CO_2$ emissions, the Energy Performance of Buildings Directive requires energy retrofitting of around five of the twelve million existing residential buildings in Italy. Owners and investors require simple tools that can guide them in their decision-making, encompassing the economic aspects, from the earliest planning stages. The extensive availability of data on buildings and, in particular data from EPCs, and the proliferation of machine learning techniques have led to the diffusion of data-driven tools to inform design decisions.

Data-driven methods can be heavily influenced by the data utilised during the construction of the method itself. In this specific case, EPC data used represent the energy performance of residential buildings located in Northern Italy, and consequently, the method can be used to assess the economic feasibility of energy retrofitting interventions in this region. However, its adaptability to other areas is straightforward. By expanding the database on which the methodology is based, it would be possible to extend its effectiveness to other regions. There is nothing to prevent the use of EPCs from all over Italy or any other European country that uses the same criteria for assessing the energy performance of buildings. On the other hand, it would be difficult to use the method in countries that do not use EPCs, as it would require profound changes to the neural network that predicts energy consumption.

The tool above proposed aids owners and investors to make design choices at the earliest stages of the decision-making process. The exploitation by building owners or potential investors is possible because the tool has been designed, since its conception, for the use by individuals not acquainted with the mathematical models describing the energy performance of buildings, with those for calculating costs over the life cycle of buildings, or with the mathematical techniques required to deal with all the uncertainties that characterise the input parameters of the aforementioned models. To ascertain the economic feasibility of an intervention, only eleven input data are required, comprising of eight that depict the building's features and three that give a description of the retrofit intervention. The first eight can be readily acquired from the energy performance certificate, if available, while the three relating to the retrofit intervention may be estimated even by non-professionals. Nevertheless, the simplicity of the method and the very limited number of input parameters did not affect the reliability of the results, which remained very high, as shown by the results of the neural network training phase presented in Sect. 3.2.5. This reliability must be compared with the typical uncertainty of LCC analysis results, which can reach peaks of 37% with an average of 26% around the median (Pernetti et al. 2021).

The use of a graphical scheme resembling energy performance certificates, with similar colours, reflects a preference for simplicity in presenting the findings. Looking at Fig. 3.9, the person who is going to invest in the retrofit can read, even without any experience in the field, the probability of the investment being paid back in 15, 20 or 25 years. On the basis of this initial assessment, the investor can then decide whether to proceed to the design phase of the intervention, whether to change the type of intervention, or even to abandon the idea of retrofitting the building.

The proposed tool's rapid functionality enables its effective application to extensive building portfolios for determining retrofit policies. However, for such large-scale applications, it is necessary to forego the graphical display of payback period probabilities, as illustrated in Fig. 3.9.

In conclusion, the tool presented in this article for predicting the probability of meeting a predefined payback period allows decisions on the energy retrofitting of a building or portfolio of buildings with minimal cost and maximum impact, suitable for a wide range of users due to its simplicity and reliability.

## 3.3   Conclusions

This chapter highlights the critical importance of incorporating sustainability consideratons from the outset of construction projects. It posits that decisions made in the early design phases serve as the foundation for sustainable construction, influencing long-term outcomes in terms of environmental impact, economic viability and social benefits. The chapter highlights how early design decisions are critical in determining operational energy use. These early decisions set the trajectory for the project, making subsequent changes to improve sustainability increasingly difficult and costly as the project progresses.

The chapter explores the integration of AI and big data analytics as transformative tools for enhancing sustainability in construction. AI's capacity to analyse large and complex datasets can significantly improve decision-making processes throughout a project's lifecycle, enabling more strategic and effective interventions at the design stage. However, the successful application of AI hinges on data quality. Inaccurate or biased data can lead to flawed outputs, potentially resulting in financial losses and safety risks. Therefore, robust data governance practices are essential to harness AI's full potential in the construction industry.

The use of big data and AI extends to understanding the built environment and making strategic decisions, such as decarbonizing building portfolios. Traditional decision-making in construction often relies on expert judgments based on limited data, which can lead to inefficiencies like cost overruns and delays. In contrast, data-driven approaches ensure that decisions are grounded in concrete data, enhancing transparency, accountability, and accuracy.

The chapter also details the use of machine learning (ML) in forecasting building energy performance and assessing the economic feasibility of energy retrofits. It presents a data-driven methodology to predict the payback period of retrofitting investments, utilising Energy Performance Certificates (EPCs) to gauge energy savings and investment viability. The methodology's adaptability to other regions is noted, provided that compatible EPC data is available.

In summary, this chapter highlights the transformative potential of AI and big data in promoting sustainability in the construction sector. By enabling more informed and strategic decision-making earlier in the project lifecycle, these technologies can significantly reduce the environmental impact of construction activities and support the industry's transition to more sustainable practices. Incorporating data-driven approaches in the early stages of design not only enhances the architectural quality of buildings, but also contributes to the broader goals of reducing carbon emissions and promoting long-term economic and social sustainability.

# References

ARIA. CENED—Certificazione ENergetica degli EDifici (2018), https://www.dati.lombardia.it/res ource/rsg3-xhvk.json

E. Baldoni, S. Coderoni, E. Di Giuseppe, M. D'Orazio, R. Esposti, G. Maracchini, A software tool for a stochastic life cycle assessment and costing of buildings' energy efficiency measures. Sustainability 13(14) (2021). https://doi.org/10.3390/su13147975

S.E. Bibri, J. Krogstie, The emerging data–driven smart city and its innovative applied solutions for sustainability: the cases of London and Barcelona. Energy Inf. 3(1), 5 (2020). https://doi.org/10.1186/s42162-020-00108-6

L. Bragança, S.M. Vieira, J.B. Andrade, Early stage design decisions: the way to achieve sustainable buildings at lower costs. Sci. World J. 2014, 365364 (2014). https://doi.org/10.1155/2014/365364

E. Brynjolfsson, L.M. Hitt, H.H. Kim, Strength in numbers: how does data-driven decisionmaking affect firm performance? SSRN Electron. J. (2011). https://doi.org/10.2139/ssrn.1819486

M.E. Celebi, K. Aydin, Unsupervised Learning Algorithms, vol. 9 (Springer, 2016)

P. Chastas, T. Theodosiou, D. Bikas, Embodied energy in residential buildings-towards the nearly zero energy building: a literature review. Build. Environ. 105, 267–282 (2016). https://doi.org/10.1016/j.buildenv.2016.05.040

P.-Y. Chen, X. Guan, A multi-source data-driven approach for evaluating the seismic response of non-ductile reinforced concrete moment frames. Eng. Struct. 278, 115452 (2023). https://doi.org/10.1016/j.engstruct.2022.115452

Q. Cheng, R. Oberhänsli, M. Zhao, A new international initiative for facilitating data-driven earth science transformation. Geol. Soc. Lond. Spec. Publ. 499(1), 225–240 (2020). https://doi.org/10.1144/SP499-2019-158

A.K. Dubey, V. Jain, Comparative study of convolution neural network's ReLU and Leaky-ReLU activation functions, in Applications of Computing, Automation and Wireless Systems in Electrical Engineering. ed. by S. Mishra, Y.R. Sood, A. Tomar (Springer, Singapore, 2019), pp.873–880

ENEA—National agency for new technologies, energy and sustainable economic development. SUPERBONUS 110 statistics, 31 Oct 2023 (2023), https://www.efficienzaenergetica.enea.it/component/jdownloads/?task=download.send&id=593&catid=40&Itemid=593

European Commission. A clean planet for all a European strategic long-term vision for a prosperous, modern, competitive and climate neutral economy (2018), https://eur-lex.europa.eu/legal-content/EN/TXT/?uri=CELEX:52018DC0773

European Parliament. Directive (EU) 2024/1275 of the European Parliament and of the Council of 24 April 2024 on the energy performance of buildings (recast) (2024), http://data.europa.eu/eli/dir/2024/1275/oj

E. Favarelli, E. Testi, A. Giorgetti, The impact of sensing parameters on data management and anomaly detection in structural health monitoring. J. Civ. Struct. Heal. Monit. 12(6), 1413–1425 (2022). https://doi.org/10.1007/s13349-022-00566-4

M. Feria, M. Amado, Architectural design: sustainability in the decision-making process. Buildings 9(5), 135 (2019). https://doi.org/10.3390/buildings9050135

L. Gabrielli, A.G. Ruggeri, Developing a model for energy retrofit in large building portfolios: energy assessment, optimization and uncertainty. Energy Build. 202, 109356 (2019). https://doi.org/10.1016/j.enbuild.2019.109356

A. Galimshina, M. Moustapha, A. Hollberg, P. Padey, S. Lasvaux, B. Sudret, G. Habert, Statistical method to identify robust building renovation choices for environmental and economic performance. Build. Environ. 183, 107143 (2020). https://doi.org/10.1016/j.buildenv.2020.107143

M. Ghafoori, M. Abdallah, Innovative optimization model for planning upgrade and maintenance interventions for buildings. J. Perform. Constr. Facil. 36(6) (2022). https://doi.org/10.1061/(asce)cf.1943-5509.0001757

E. Di Giuseppe, A. Massi, M. D'Orazio, Impacts of uncertainties in life cycle cost analysis of buildings energy efficiency measures: application to a case study. Energy Procedia **111**, 442–451 (2017). https://doi.org/10.1016/j.egypro.2017.03.206

Y.M. Goh, L.B. Newnes, A.R. Mileham, C.A. McMahon, M.E. Saravi, Uncertainty in through-life costing—review and perspectives. IEEE Trans. Eng. Manag. **57**(4), 689–701 (2010). https://doi.org/10.1109/TEM.2010.2040745

G. Grazieschi, P. Gori, L. Lombardi, F. Asdrubali, Life cycle energy minimization of autonomous buildings. J. Build. Eng. **30**, 101229 (2020). https://doi.org/10.1016/j.jobe.2020.101229

M. Greenacre, P.J.F. Groenen, T. Hastie, A.I. D'Enza, A. Markos, E. Tuzhilina, Principal component analysis. Nat. Rev. Methods Prim. **2**(1), 100 (2022). https://doi.org/10.1038/s43586-022-00184-w

S. Hansen, F. Fassa, S. Wijaya, Factors influencing scheduling activities of construction projects. J. Leg. Aff. Dispute Resolut. Eng. Constr. **15**(1) (2023). https://doi.org/10.1061/(ASCE)LA.1943-4170.0000594

A. Hardy, D. Glew, An analysis of errors in the Energy Performance certificate database. Energy Policy **129**, 1168–1178 (2019). https://doi.org/10.1016/j.enpol.2019.03.022

X. Jn, F. Xiao, C. Zhang, Z. Chen, Semi-supervised learning based framework for urban level building electricity consumption prediction. Appl. Energy **328**, 120210 (2022). https://doi.org/10.1016/j.apenergy.2022.120210

S. Kaewunruen, Q. Lian, Digital twin aided sustainability-based lifecycle management for railway turnout systems. J. Cleaner Prod. **228**, 1537–1551 (2019). https://doi.org/10.1016/j.jclepro.2019.04.156

D.P. Kingma, J. Ba, Adam: a method for stochastic optimization, in *3rd International Conference for Learning Representations*, 22 December 2017 (2017), http://arxiv.org/abs/1412.6980

M. Köppen, The curse of dimensionality, in *5th Online World Conference on Soft Computing in Industrial Applications (WSC5)*, vol. 1 (2000), pp. 4–8

Y. Lai, S. Papadopoulos, F. Fuerst, G. Pivo, J. Sagi, C.E. Kontokosta, Building retrofit hurdle rates and risk aversion in energy efficiency investments. Appl. Energy **306**, 118048 (2022). https://doi.org/10.1016/j.apenergy.2021.118048

D. Lee, S.H. Lee, N. Masoud, M.S. Krishnan, V.C. Li, Integrated digital twin and blockchain framework to support accountable information sharing in construction projects. Autom. Const. **127**, 103688 (2021)

R. Lu, I. Brilakis, Digital twinning of existing reinforced concrete bridges from labelled point clusters. Autom. Constr. **105**, 102837 (2019). https://doi.org/10.1016/j.autcon.2019.102837

T. Márk, S. Ádám, Benefits of transforming organizations to be data-driven and more innovative (2023), https://controllerinfo.hu/wp-content/uploads/2023/12/ContrInf_beliv_kulonszam_2023-7.pdf

J. Moolayil, Keras in action, in *Learn Keras for Deep Neural Networks: A Fast-Track Approach to Modern Deep Learning with Python*, ed. by J. Moolayil (Apress, 2019), pp. 17–52. https://doi.org/10.1007/978-1-4842-4240-7_2

D. Palladino, S. Di Turi, Energy and economic savings assessment of energy refurbishment actions in Italian residential buildings: comparison between asset and tailored calculation. Sustainability **15**(4) (2023). https://doi.org/10.3390/su15043647

R. Pernetti, F. Garzia, U. Filippi Oberegger, Sensitivity analysis as support for reliable life cycle cost evaluation applied to eleven nearly zero-energy buildings in Europe. Sustain. Cities Soc. **74**, 103139 (2021). https://doi.org/10.1016/j.scs.2021.103139

H.N. Rafsanjani, A.H. Nabizadeh, Towards digital architecture engineering and construction (AEC) industry through virtual design and construction (VDC) and digital twin. Energy Built Environ. **4**(2), 169–178 (2023). https://doi.org/10.1016/j.enbenv.2021.10.004

T. Ramesh, R. Prakash, K.K. Shukla, Life cycle energy analysis of buildings: an overview. Energy Build. **42**(10), 1592–1600 (2010). https://doi.org/10.1016/j.enbuild.2010.05.007

F. Re Cecconi, A. Khodabakhshian, L. Rampini, Data-driven decision support system for building stocks energy retrofit policy. J Build. Eng. **54**, 104633 (2022). https://doi.org/10.1016/j.jobe. 2022.104633

L.B. Robichaud, V.S. Anantatmula, Greening project management practices for sustainable construction. J. Manag. Eng. **27**(1), 48–57 (2011). https://doi.org/10.1061/(ASCE)ME.1943-5479.0000030

B. Schölkopf, A. Smola, K.-R. Müller, Kernel principal component analysis, in *Artificial Neural Networks — ICANN'97*. ed. by W. Gerstner, A. Germond, M. Hasler, J.-D. Nicoud (Springer, Berlin, 1997), pp.583–588

S. Senaratne, S. Farhan, Role of standard contracts in mitigating disputes in construction. J. Leg Aff. Dispute Resolut. Eng. Constr. **15**(1) (2023). https://doi.org/10.1061/(ASCE)LA.1943-4170.000 0593

M.A. Syakur, B.K. Khotimah, E.M.S. Rochman, B.D. Satoto, Integration k-means clustering method and elbow method for identification of the best customer profile cluster. IOP Conf. Ser.: Mater. Sci. Eng. **336**, 12017 (2018)

J. Terés-Zubiaga, I. González-Pino, I. Álvarez-González, Á. Campos-Celador, Multidimensional procedure for mapping and monitoring urban energy vulnerability at regional level using public data: proposal and implementation into a case study in Spain. Sustain. Cities Soc. **89**, 104301 (2023). https://doi.org/10.1016/j.scs.2022.104301

United Nations Environment Programme, Global status report for buildings and construction: towards a zero-emission, efficient and resilient buildings and construction sector (2022), https://www.unep.org/resources/publication/2022-global-status-report-buildings-and-construction

S. Wan, S. Guan, Y. Tang, Advancing bridge structural health monitoring: insights into knowledge-driven and data-driven approaches. J. Data Sci. Intell. Syst. (2023). https://doi.org/10.47852/bonviewJDSIS3202964

L. Wederhake, S. Wenninger, C. Wiethe, G. Fridgen, On the surplus accuracy of data-driven energy quantification methods in the residential sector. Energy Inf. **5**(1), 7 (2022). https://doi.org/10.1186/s42162-022-00194-8

Z.F. Wu, J. Li, M.Y. Cai, Y. Lin, W.J. Zhang, On membership of black-box or white-box of artificial neural network models, in *2016 IEEE 11th Conference on Industrial Electronics and Applications (ICIEA)* (2016), pp. 1400–1404. https://doi.org/10.1109/ICIEA.2016.7603804

H. Xie, M. Xin, C. Lu, J. Xu, Knowledge map and forecast of digital twin in the construction industry: State-of-the-art review using scientometric analysis. J. Cleaner Prod. **383** (2023). https://doi.org/10.1016/j.jclepro.2022.135231

R. Xu, D. Wunsch, Survey of clustering algorithms. IEEE Trans. Neural Netw. **16**(3), 645–678 (2005)

A.F.M. Zammari, M.F. Ayob, Data quality issues that hinder the implementation of artificial neural network (ANN) for cost estimation of construction projects in Malaysia. J. Archit. Plan. Constr. Manag. **13**(1), 40–53 (2023). https://doi.org/10.31436/japcm.v13i1.731

X. Zhang, S. Huang, S. Yang, R. Tu, L. Jin, Safety assessment in road construction work system based on group AHP-PCA. Math. Probl. Eng. **2020**, 1–12 (2020). https://doi.org/10.1155/2020/6210569

# Chapter 4
# AI for Construction Risk Management

Project Risk Management aims to exploit or enhance positive risks (opportunities) while avoiding or mitigating negative risks (threats) through seven main steps: (a) Plan Risk Management, (b) Identify Risks, (c) Perform Qualitative Risk Analysis, (d) Perform Quantitative Risk Analysis, (e) Plan Risk Responses, (f) Implement Risk Responses, (g) Monitor Risks (Project Management Institute (PMI) 2017). RM is an important tactic to meet project targets like time, budget, and quality (Han et al. 2008). However, conventional RM is conducted in an inefficient, subjective, and superficial form and based on individual and experience-based expert judgments (Li et al. 2018). Furthermore, the data registry is not done in a structured, interoperable, and regular manner. Therefore, knowledge transfer, model generalization, and process automation remain critical issues for future projects' RM (Eybpoosh et al. 2011).

The advancement of AI and digital technologies can significantly change conventional risk assessment and management methods, making them factual, efficient, generalizable, and able to be performed in real-time (Chenya 2022). AI models can improve analytical capabilities across the RM domain while offering a high granularity and depth of predictive analysis (Guzman-Urbina et al. 2018), and provide accurate results in uncertain, dynamic, and complex environments (Yaseen et al. 2020), like the construction industry. AI-based RM systems can function as (a) early warning systems for risk control, (b) AI-based risk analysis systems using algorithms such as neural networks for identifying complex data patterns, (c) risk-informed decision support systems for predicting various outcomes and scenarios of the decisions, (d) game-theory-based risk analysis systems, (e) data mining systems for large data sets, (f) agent-based RM systems for supply chain management risks, (g) engineering risk analysis systems based on optimization tools, and (h) knowledge management systems by integrating decision sup-port systems, AI, and expert systems, to capture

By Ania Khodabakhshian.

© The Author(s), under exclusive license to Springer Nature Switzerland AG 2024
F. Re Cecconi et al., *Building Tomorrow: Unleashing the Potential of Artificial Intelligence in Construction*, PoliMI SpringerBriefs,
https://doi.org/10.1007/978-3-031-77197-2_4

the tacit knowledge within organizations' computer systems (Wu, Chen and Olson 2014).

Machine Learning, a branch of AI, combines methods from statistics, database analysis, data mining, pattern recognition, and AI to extract trends, interrelationships, patterns of interest, and valuable insights from complex data sets (Flath et al. 2012). ML techniques have been widely studied in construction RM research, aiding in hazard and risk identification, vulnerability assessment, consequence prediction, and mitigation strategy development (Habibi Rad et al. 2021), which can bring numerous benefits to construction projects, including preventing cost overruns, enhancing site safety, and managing projects efficiently (Regona et al. 2022). However, RM is a lesser-studied and progressed domain in construction projects due to the complex and probabilistic nature of assessments, inferences, and the direct influence of RM on other knowledge areas, such as stakeholder management (Xia et al. 2018). The key reasons are:

(a)  Lack of structured data and infrequent documentation in the projects
(b)  Over-reliance on individual and experience-based judgment by experts in RM
(c)  Isolated risk analysis and ignorance of the causal inferences between variables in risk path analysis, and
(d)  Incorrect choice of the AI model for a given problem, regarding data availability and requirements, the role of probability, expert judgement, and the reasoning behind the analysis (An et al. 2021; Chenya 2022).

AI and specifically ML, can are applied in various domains relevant to construction RM, such as:

(a)  Safety Management covers four main areas: (a) Analyzing factors that affect the safety performance in projects (Chan et al. 2017), (b) Selecting Safety Management Strategies and Interventions (Mofidi et al. 2020), (c) Safety Supervision and workforce monitoring (Nath, Behzadan and Paal 2020), (d) Other Safety-Related Topics, including lifecycle safety control, safety design, and accident diagnosis (Abdat et al. 2014).
(b)  Risk Management in building, infrastructure, excavation projects, and energy (e.g., buildings, bridges, tunnels, power plants) (Wang et al. 2014).
(c)  Contract and Procurement Management is used to analyze construction contractual risks, handle disputes, and improve bidding (Abotaleb and El-adaway 2017). Particularly, the Naïve Bayes method has been employed to extract the required contractual text for decision-making (Hassan and Le 2020).
(d)  Process Control for managing project schedules, predicting schedule performance, productivity management, and other areas like progress monitoring and performance measurement (Golparvar-Fard et al. 2015; Sabillon et al. 2020)
(e)  Project Cost Management, for cost prediction, forecasting errors in cost estimation, and dynamic monitoring of construction costs (Nasrazadani et al. 2017).
(f)  Quality Management, including evaluating the impacts of stakeholders on quality defects, evaluating operator welding-quality performance, and examining building materials compliance for fire safety (Yu et al. 2019).

(g) Other CM Research and Practice, such as design management, project information management, environment management, materials management, and stakeholder management (Hu and Castro-Lacouture 2019).

In this section, the different ML algorithms applications for construction risk identification, assessment, and control, which previous researchers studied, are listed and analyzed, such as Artificial Neural Networks (Heravi, Asce and Eslamdoost 2015), Decision Trees (Chou and Lin 2013), Logistic Regression (Hwang and Kim 2016), Naïve Bayesian Models (Gerassis et al. 2017), Support Vector Machines (Huang and Tserng 2018), Genetic Algorithm (GA), Structure Equation Modelling (SEM), Fuzzy Hybrid Methods (FHMs) (Afzal et al. 2019). Followingly, a case-based comparison between some of these algorithms is conducted to assess the effect of database size, data quality and quantity, type of problem, and probabilistic risk reasoning on the performance of each of the methods.

## 4.1 Phase-Based Classification of AI Applications

Construction Risk identification has been conducted by various methods such as construction drawing, meta-network, Monte Carlo simulation, ontology, and BNs (Liu et al. 2021). Moreover, various techniques have been used to model the interdependencies of project risks in literature, including Structural Equation Modeling (SEM) (Eybpoosh et al. 2011), Analytic Network Process (ANP) (Prince Boatenga and Ogunlana 2015), causal mapping (Ackermann and Alexander 2016) systems thinking (Loosemore and Cheung 2015), and Bayesian Belief Networks (BBNs) (Yildiz et al. 2014), among which BBNs have gained much popularity due to benefiting from a robust theoretical framework and the ability to capture uncertainty and update beliefs upon the availability of new information, which is a considerable advantage in ongoing projects. Qualitative Risk Analysis is the process of prioritizing individual project risks for further analysis or action by assessing their probability of occurrence and impact as well as other characteristics (Project Management Institute (PMI) 2017). Various AI techniques such as multilevel regression (SEM) (Ebrat and Ghodsi 2014), MCDM, probability models, FHM, NNs, and genetic algorithm (GA) have been used in previous research for both qualitative and quantitative analysis. Quantitative Risk Analysis is the process of numerically analyzing the combined effect of identified individual project risks and other sources of uncertainty on overall project objectives like cost and schedule. Quantitative Risk assessment tools are based on different linear and non-linear approaches. However, since construction projects have stochastic behavior, non-linear probabilistic models of AI, such as ANNs, ANP, and BBN, are dominant to address this phenomenon of interdependency (Dekker 2013), which can be used independently or in a hybrid manner. Few studies have adopted hybrid methods based on the FST and other AI tools to design flexible risk assessment tools under high uncertainty (Afzal et al. 2019).

## 4.2   AI Applications Classifications in Literature

Various categories have been proposed for AI-based Risk analysis and reasoning methods in the literature. Based on the categorization for AI application areas in the construction industry proposed by Pan and Zhang (2021), RM falls under (a) the category of Expert Systems/Fuzzy logic for Knowledge Representation and Reasoning mainly formed on probabilistic, qualitative, and linguistic analysis, and (b) Machine Learning for supervised learning based on either probabilistic or deterministic analysis. Samantra et al. (2017) classified construction risk assessment approaches as (a) Probabilistic approach, dealing with risk probability and impact estimation based on historical numeric data, including Sensitivity analysis, Decision Tree analysis, Bayesian Networks, Monte Carlo simulation, etc. (Zhang et al. 2014a, b), and (b) Possibilistic approach, dealing with risk probability and impact estimation based on qualitative or descriptive data including fuzzy logic (Dikmen et al. 2007). The advantage of possibilistic models is that they can embrace the uncertain and vague definition of risk factors and their magnitude in a linguistic and subjective human description (Samantra et al. 2017). Although called by various names, the notion and reasonings for classifying all of the methods are the same. For ease of reference, this research calls them Probabilistic and Deterministic models. ML algorithms can generally conduct deterministic or probabilistic analyses, which are grouped under deterministic or probabilistic approaches (Khodabakhshian et al. 2023). It is noteworthy that this classification basis is the risk reasoning itself, which is applicable to all phases of the RM process, from risk identification to assessment and mitigation planning.

Probabilistic models are mostly based on Bayesian Inference, which allows making judgments on prior and posterior probabilities in random variables based on various sources, like expert judgment, model simulation, or historical data (Debnath et al. 2016). Prior probability is the likelihood of a particular state of a variable happening without seeing any evidence, and posterior probability is the updated belief or likelihood of that state of a variable happening after seeing evidence (Zhang et al. 2016). Benefitting from multiple sources of data in probabilistic approaches, the priors can be learned based on one source and the posteriors can be updated by another source. This is a huge advantage in situations with limited data, as the application of multiple sources compromises the data limitation. Moreover, they provide a probability distribution of possible outcomes, considering the interrelation and causal inferences of input variables on each other. Therefore, they do not need an extensive database to draw judgment from and can update the probability distribution based on new observations or data (Gelman et al. 2013). The downside of the Probabilistic approaches, on the other hand, is the subjectivity, bias, and overreliance on experts' opinions if not appropriately calibrated (Bar-hillel and Neter 1993). These methods have a vast application in expert systems and knowledge representation and can have one of the aforementioned risk reasonings (Wee et al. 2015): (a) Probability-based reasoning refers to probability theory to indicate the uncertainty in knowledge, including fault tree analysis (FTA), SEM, and BNs. Figures 5.1 and 5.2 present the structures of a fault tree and an event tree, (b) Rule-based reasoning, deploying a set of rules in the "if <conditions>, then <conclusion>" format with logical connectives, like AND, OR, NOT, for analysing qualitative and linguistic data of expert opinion, including Fuzzy Logic, (c) Fuzzy Cognitive Map (FCM) learned from data or expert opinions, in which the fuzzy graph structure enables interpreting complex relationships and systematic causal propagation for immediate identification of risks' root causes in uncertain conditions. Figure 5.3 presents the structure of an FCM.

On the other hand, Deterministic models are mostly based on the Frequentist approach, which can be merely based on historical records, and the priors are learned based on the frequency of an event happening in the database provide a definite prediction of output value without assigning a probability distribution to it, which is their main difference from Probabilistic models. These methods perform best when a huge amount of data is available, capturing linear and nonlinear relationships of the data and serving as a predicting model for industrial RM control and accident severity assessment (Gondia et al. 2020a). The learning and development processes are much more straightforward and simpler compared to the Probabilistic methods, as the elicitation process to obtain information on probabilities from experts is usually challenging and time-consuming. However, the downside, in contrast to Probabilistic approaches, is the inability to assign probability to a particular event happening after witnessing evidence, i.e., the posterior update. Deterministic Models include most of the ML algorithms, including (a) Regression to predict continuous numerical outcomes like delay caused by a risk, including Linear Regression, Decision Trees, Support Vector Machines (SVM), and Neural Networks (NN) techniques, (b) Classification to present the class of the output based on some input features like risk identification including NNs, Random Forest, SVM, and Genetic Algorithm, (c) Clustering to explore data for natural groupings like finding related events causing a risk including K-means and SVM, (d) Attribute importance to rank attributes based on their relationships to the target variable like identifying the most significant causes of accidents including Decision Trees and Random Forest, (e) detection to identify unusual cases based on deviation like identifying accident risks including SVM and Deep Neural Networks (Ajayi et al. 2019).

Some of the most common ML-based techniques in previous literature include.

## 4.3  Theoretical Comparison of Probabilistic and Deterministic ML Models for RM

In general, algorithms with a deterministic approach have advanced structure, quicker processing time, and higher precision of results in complex problems. However, they require a large amount of structured data with no missing values or uncertainties. Given that documentation is in a non-optimum condition in the industry, data scarcity and infrequent data updates are the main challenges in these models. The probabilistic approach, on the other hand, is more appropriate for RM in construction due to functioning in the state of data scarcity and missing values and being closer to reality, considering the inter-dependencies between risk variables. They can integrate subjective and experience-based experts' opinions through elicitation with objective historical data gathered from previous projects or simulations to overcome the data scarcity issue. Moreover, they benefit from the risk path approach instead of isolated risk assessment, which makes the assessment process closer to reality. However, the structure and parameter learning are daunting and complicated tasks as the model becomes more complex, containing more variables and risk factors. Table 4.1 compares these two categories of algorithms based on a plethora of criteria, such as reasoning basis, data source, inference, and structure, which can be a good benchmark for the practical application of each algorithm based on the conditions of the problem.

## 4.4  Practical Comparison of Probabilistic and Deterministic ML Models for RM

In order to compare the performance of the probabilistic and deterministic ML models in identifying and evaluating risks in construction projects, various algorithms from each category were applied to two case studies and different sizes. The main objective of implementing the proposed solutions on two completely different databases was to conduct a comparative analysis between the results to delineate the importance of the database size and type of data in the performance of each ML algorithm. As a result, this research can contribute to construction companies' proper choice of ML algorithms with respect to their data availability.

The first case study is a small database of 44 construction projects in Italy, containing 46 columns as the effective project variables, and a list of 65 common risks as the target variables, some of which happened in these projects and their values are equal to 1 in the database. It is noteworthy that the main focus of this case study was the project risks, hence the effective variables were also on project level, such as initial budget and schedule, project delivery method, contract type, number of contractor, and the type of the project. Figure 4.1 presents the relationship between the value of different features in the database and the Mean Absolute Error (MAE) of the prediction of each project or row of data; where each line represents a different project, the x-axis represents the different project features, including the MAE, and the y-axis represents the standardized values of these features.

The second case study instead, is a is an open-access database sourced from the Capital Project Schedules and Budgets database available on the City of New York's Open Data Portal (https://data.cityofnewyork.us/Housing-Development/Capital-Project-Schedules-andBudgets/2xh6-psuq), maintained by the New York City government. It includes 13,570 rows or projects and 14 columns or project attributes, including Project Geographic District and Building Identifiers, funding source,

**Table 4.1**  Analytical comparison between probabilistic and deterministic RM models (Khodabakhshian et al. 2023)

| Comparison criteria | Probabilistic approach | Deterministic approach |
| --- | --- | --- |
| Reasoning basis | Probability-based reasoning Rule-based reasoning Fuzzy logic (Samantra, Datta and Mahapatra 2017; Valpeters, Kireev and Ivanov 2018) | Forward propagation and backpropagation Loss function Weights and biases (Hosny et al. 2015; Habbal et al. 2020) |
| Structure | Interconnected graphs (Khakzad et al. 2013; Qazi et al. 2016; Lee and Kim 2017) | Layers of neurons or branches (Jin and Zhang 2011; Gajzler 2013) |
| Data source | Historical Data, model simulation Experts' opinion (Butler et al. 2015; Mkrtchyan et al. 2015) | Historical data, model simulation (Hosny et al. 2015; Habbal et al. 2020; Re Cecconi et al. 2022) |
| Inference | Bayesian inference (Nguyen and Tran 2016) | Frequentist inference (Lele and Allen 2006) |
| Data requirements | Limited amount of data Able to deal with missing values Numerical, categorical, and linguistic data (Regan et al. 2002; Mohamed and Tran 2021) | High amount of data Partial ability to deal with missing values (Fan et al. 2019) |
| Probability and dependencies' role | Embrace probability in assessments Considering variables interdependencies with each other and final output (Omondi et al. 2021; Wang et al. 2021) | Does not embrace probability in assessments Considering variables interdependencies on final output (Valpeters et al. 2018; Anysz et al. 2021) |
| Prediction precision | Mid-high (Tardioli et al. 2020) | Very high (Akinosho et al. 2020) |
| Application scope | Subjective and uncertain problems with limited data (Yang et al. 2008) | Objective and complex problems with abundant data (Gondia et al. 2020b) |
| Application in RM processes | Risk identification Qualitative analysis Risk control (Karakas et al. 2013; Islam et al. 2019; Yucelgazi and Yitmen 2020) | Risk identification Qualitative and quantitative analysis Mitigation planning Risk control (Fang and Marle 2013; Chattapadhyay et al. 2021) |

(continued)

**Table 4.1** (continued)

| Comparison criteria | Probabilistic approach | Deterministic approach |
|---|---|---|
| Advantages | Flexibility to various problems Ability to integrate qualitative and quantitative data (subjective and objective) Risk path approach Ability to include dynamic data (Serpella et al. 2014; Zhang et al. 2014b) | Quick processing and learning Ability to consider linear and nonlinear relationships among data Ability to include dynamic data (Sherafat et al. 2020; Von Platten et al. 2020) |
| Disadvantages | Takes longer time to create the structure Not high precision if merely based on historical data High processing time in complex problems (Wisse et al. 2008; Qazi et al. 2016) | Individual risk analysis approach (isolated) Not flexible toward change Requirement of high data volume (Giannakos and Xenidis 2018; Lamine et al. 2020) |

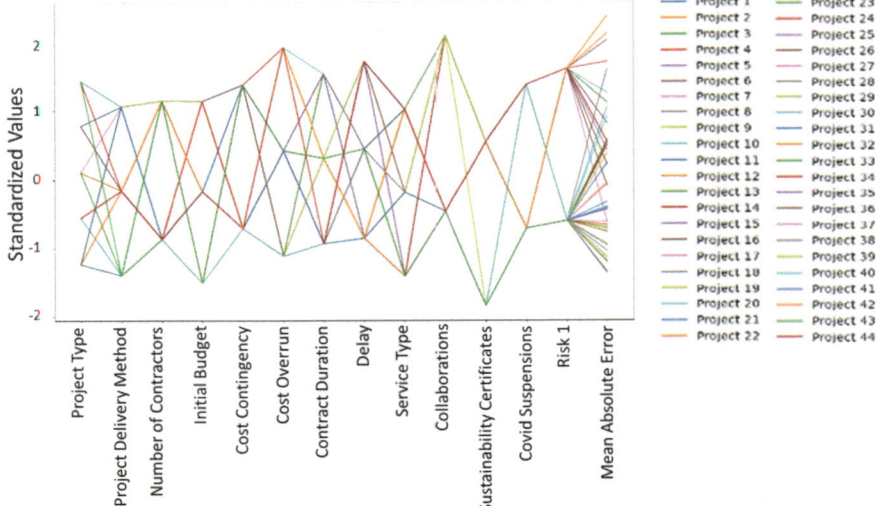

**Fig. 4.1** Relationship between project features and MAE value for each project for risk 1 prediction

Project Description (Description of construction/ retrofit services and work packages), Project Phase Name and Status (completed, ongoing), Project planes and actual start and finish dates, and project initial budget and final total cost. Table 4.2 presents the statistical information of the database.

The research methodology, as briefly described in Fig. 4.2, involves the development of three types of models based on three different combinations of data sources, namely, (a) Bayesian Network based on both subjective expert data form interviews

**Table 4.2**  Statistical information of the database

| Feature | Mean | Variance | STD | Min | Max |
|---|---|---|---|---|---|
| Project geographic district | 17.67 | 83.96 | 9.16 | 1 | 32 |
| Project phase name | 0.68 | 0.55 | 0.74 | 0 | 3 |
| Week duration | 36.10 | 1163.56 | 34.11 | 0 | 314.8557 |
| Week delays | 9.53 | 528.422 | 22.98 | -24 | 171 |
| Project budget amount | 453,354.45 | 1,735,528,285,069 | 1,317,394.50 | 112 | 16,096,500 |
| Final estimate of actual costs through the end of the phase | 373,142.53 | 1,230,812,537,107 | 1,109,419.91 | 90 | 15,120,360 |
| Total phase actual spending | 335,434.41 | 984,566,654,594 | 992,253.32 | 90 | 13,610,170 |

and object project database, (b) Fuzzy Logic model based on merely expert data, and (c) Deterministic ML models based on merely objective project data. This show also the capabilities of each of the models in integrating various sources of data to based conclusions on for a more accurate and realistic prediction of risks.

In the first case study, the main challenge was the limited size of the database that was hindering the application of deterministic models with frequentist statistics. Not being to base the judgement merely on the objective project data, two solutions were applied on a probabilistic Bayesian Network model developed based on the database. First, the elicitation of the subjective experts' judgement, where 16 experts in the company were interviewed during three phases. In the first phase, they

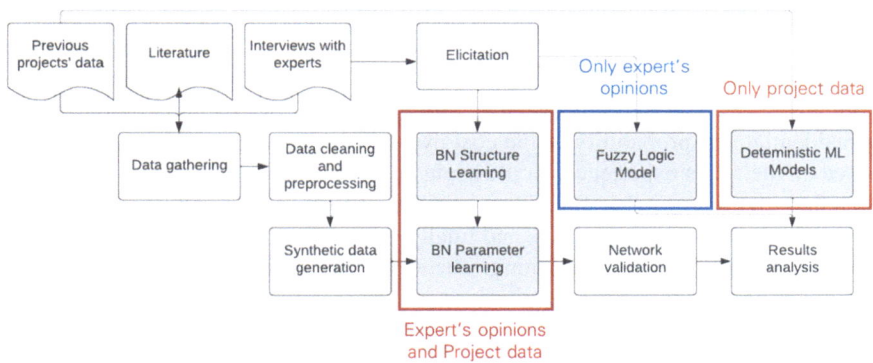

**Fig. 4.2**  Research Methodology for practical comparison of different ML models' performance in predicting risks

were asked to merely select the variables that affected each risk, with a yes or no answer, depicting the structure of the network. In the second phase, they were asked to qualitatively select the probability of each risk in the network happening given each state of the project variables, forming the parameter learning process of the BN model. And finally, in the third phase, they were asked to validate the created network with respect to their ongoing projects. Afterwards, the objective data in the database was inserted into the model and the beliefs were updated accordingly. This elicitation-based approached enabled the model to draw conclusions based on two sources of data instead of one, which compensated greatly the data limitation. And the second solution was generating synthetic data, also known as data augmentation, by Generative Adversarial Networks (GANs), which doubled the size of the database. This approach not only greatly enhanced the accuracy of the BN model, but it also enabled application of deterministic models with mere reference to the objective data. Another advantage of this solution is balancing the database and filling the gap where data is rare regarding a specific type of risk. Finally, the model was validated using the k-fold cross-validation for different risks, indicating the positive effect of data augmentation solution to increase the overall accuracy of the model from 67 to 86%.

Figure 4.3 presents the BN created for one of the risk categories, the technical risks, after structure and parameter learning based on both subjective expert data and objective project database. It indicated the overall risk exposure of the company based on their previous experiences and records, which is an effective way to translate the experience-based and individual judgement of experts into quantitative values learnable by the machine and generalizable to the next projects. After setting evidence based on a new project, the posterior probabilities of the risk nodes indicate the probability of each of the risks happening. Hence, setting a threshold of 50%, it is possible to automatically identify the risks given their probabilities surpass this threshold, as well as their probabilities, indicating the qualitative assessment of the risks. Figure 4.4 indicates the comparison of model accuracy in predicting the risks before and after data augmentation.

As a second probabilistic model, a Fuzzy Logic model was developed merely on the experts' elicitation, where all 16 experts were asked to rate the probability of the risks happening given different conditions or status of each project variable using a linguistic Likert scale. For instance, an expert could evaluate that if the project is residential, the probability of the cost overrun risk is high, while another expert, based on their own experience, would evaluate it low. All these linguistic terms were fuzzified into membership functions, aggregated to form one collective evaluation of the risk based on a set of rules, and finally, defuzzified into crip values indicating the probability of the risk happening. The ability of FL in dealing with uncertainty and imprecision in using the fuzzy values and degree of membership between 0 and 1 instead of crisp and deterministic values, enables a more flexible and nuanced representation of subjective and vague information like linguistic terms. Hence, it is easier understood by experts and can translate their experience-based subjective judgement into numerical values interpretable by the machine and generalizable to

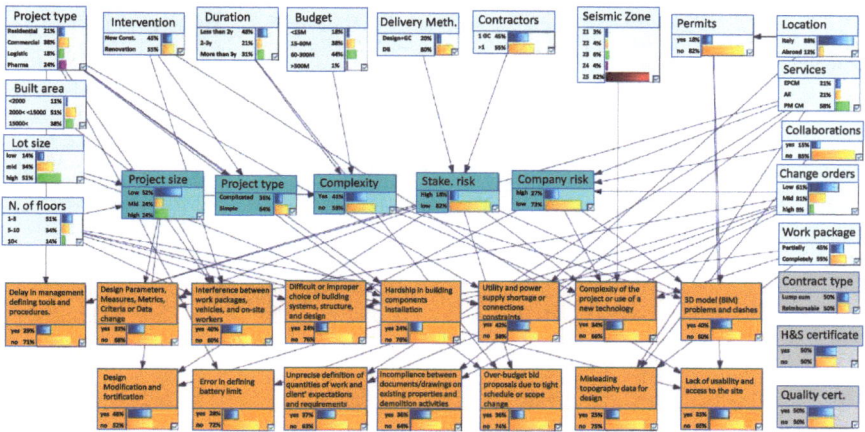

**Fig. 4.3** Technical risk network after posterior update by historical data

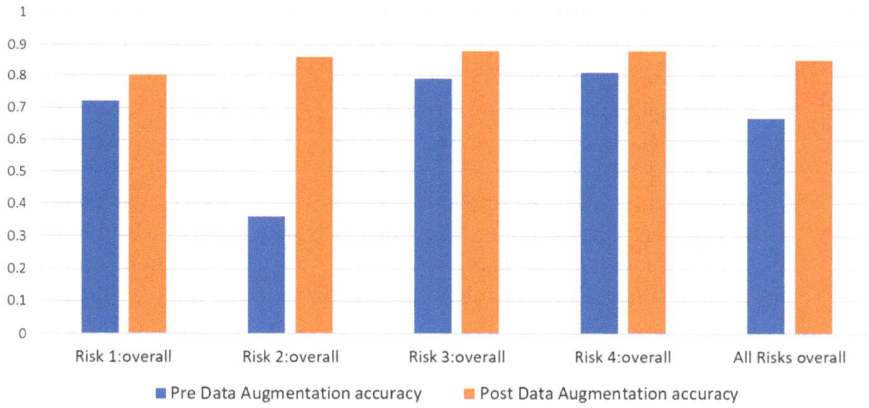

**Fig. 4.4** Comparison the pre and post data augmentation results precision of the procurement risks BN model

the next projects. Figure 4.5 indicates the 3-D rule-based reasoning modelled in MATLAB for indicating the probability distribution of the risks.

Once the size of the database increased, it was possible to apply other deterministic ML models as well for comparison. A range of ML algorithms including (a) Artificial Neural Networks (ANN), (b) Random Forest, (c) Support Vector Machine (SVM), (d) Logistic Regression, (e) XGBoost, (f) K-Nearest Neighbour, (g) Decision Tree, and (f) Naïve Bayes Classifier were applied and the results were registered. Initially, during the learning phase, the algorithms were trained on a subset of the database, where they learned to identify patterns and relationships between input and output variables. Subsequently, in the validation phase, another subset of the data was

**Fig. 4.5** Rule-based analysis between TIC, cost contingency, and financial risk in the FL model

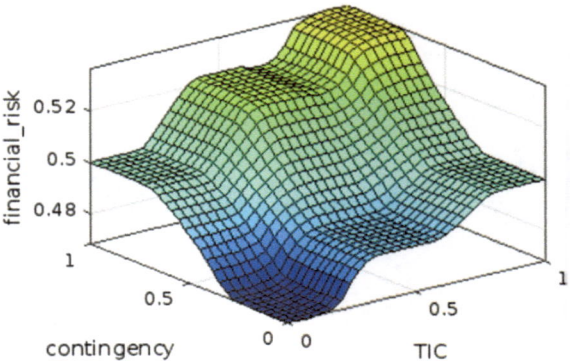

utilized to fine-tune the models' parameters. Finally, the models' performance were tested in the testing phase using a distinct dataset that had not been exposed to them during the training or validation phases. This provided an unbiased evaluation of the models' predictive accuracy and generalizability, as well as their reliability and effectiveness. Figure 4.6 indicates the comparison of the mean of the accuracy of each of these algorithms in one of the risk categories, with ANN in top. However, regardless the high precision accuracy of most of them, they have the overfitting problem given the small size of the database, hence the models are not as generalizable as the probabilistic models. Furthermore, they cannot truly predict the probability of each risk as they perform a simple classification task and for each new project, they can merely predict if the risk will or will not happen. The overall risk exposure probability in these models, however, is based on the frequentist statistic, and the more frequent a risk happened in previous projects and is more frequently repeated in the database, the models assumes it has a higher probability to happen in the future, which is not necessarily true. Deterministic models on such small databases cannot be representative enough of the reality and might undervalue the risks that historically were not so frequent in the database but might become more frequent in the future.

Moreover, the comparison of the risk probability assigned by each of the models indicated that:

(a) Judgment based on merely expert data in FL is over-conservative, estimating higher probabilities for risks.
(b) Judgment based merely on project data in deterministic ML models is not reflective of the actual situation due to the small size of the database and can underestimate the probabilities of risks. The deterministic ML models use the frequentist approach to estimate the probabilities of risks, that is, the frequency of the occurrence of risk in the database. However, if a certain risk was not repeated much in previous projects, there is no guarantee it will not be frequent in upcoming ones.
(c) Combining the two sources of judgments in BNs balances the estimates. Therefore, BNs offer the most realistic probability estimates of the risks, as evident in Fig. 4.7.

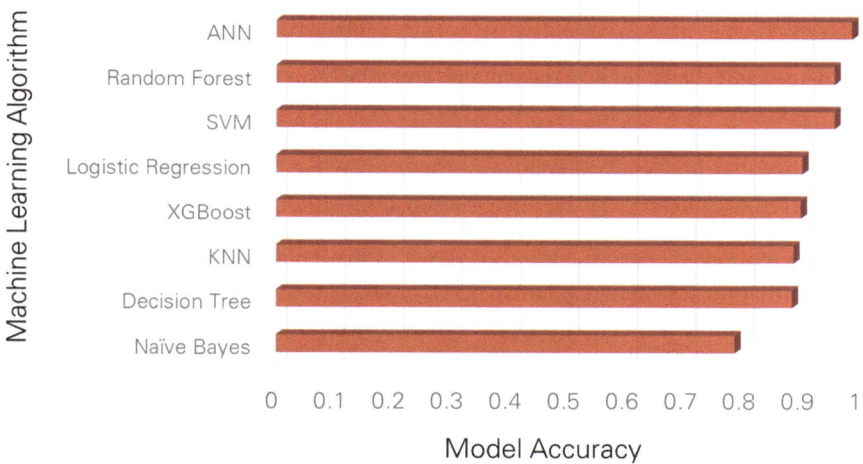

**Fig. 4.6** Comparison of the model accuracy in predicting risks for different deterministic ML algorithms

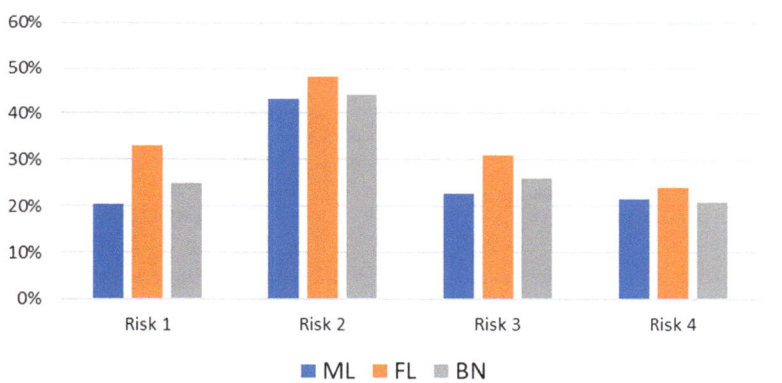

**Fig. 4.7** Comparison of risk probabilities assigned by each of the three ML, FL, and BN models

The database of the second case initially had 13,570 rows, 1489 of which remained after the data cleaning and preprocessing phases. Even after these phases, it had a massively bigger database compared to the first one, which affected the results and accuracy obtained from each algorithm. In order to compare the results, the same ML algorithms, including XGBoost, ANN, Ridge and Linear Regression, Decision Tree, and BN, were applied. However, the problem type, in contrast to the first case study, was regression. The algorithms had to predict the final delay and spending of each project.

The performance of each algorithm is assessed using the four tests of R2, Mean of Cross-Validation (CV), MSE, and MAE. Based on the obtained results, XGBoost model outperforms the other algorithms due to its robust gradient-boosting frame-work that combines multiple decision trees, enhancing accuracy and reducing overfitting. This results in the highest R-squared value (0.91), indicating a strong correlation between the predicted and observed values, as well as the lowest Mean Squared Error (45.77) and Mean Absolute Error (3.5 weeks), signifying superior prediction accuracy. Other algorithms have a good performance too, but in some cases like Decision Tree, due to its single-tree structure, are more prone to overfitting. In general, due to the huge size of the database, the black box ML models based on frequentist statistics and with more advanced structure compared to the white box probabilistic models outperform. Figures 4.8 and 4.9 presents the comparison of the performance metrics for each of the algorithms in predicting the two main risks.

In the second case study, the deterministic models outperformed the probabilistic ones, due to the availability of data and the nature of the target to predict, which was a deterministic continuous value. When enough data is available to base the judgement on, and there is no need for Bayesian inference to integrate different sources of data, the deterministic model, given their advanced structures, indicate better results. Specifically, the XGBoost model consistently demonstrated superior performance in predicting delays and total costs, followed by Decision Tree and ANN. The linear and ridge regression models exhibited lower performance compared to the non-linear models, as they assumed linear relationships between predictors and target variables,

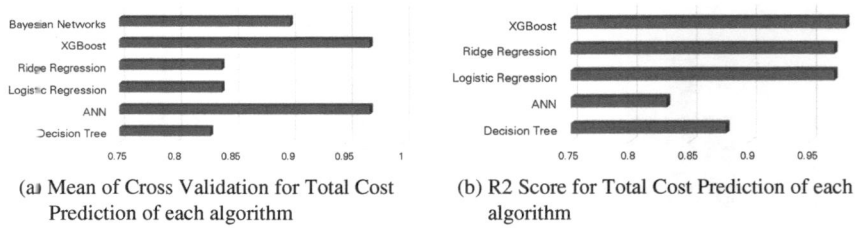

    (a) Mean of Cross Validation for Total Cost      (b) R2 Score for Total Cost Prediction of each
          Prediction of each algorithm                    algorithm

**Fig. 4.8  a** Mean of cross validation for total cost prediction of each algorithm. **b** R2 score for total cost prediction of each algorithm

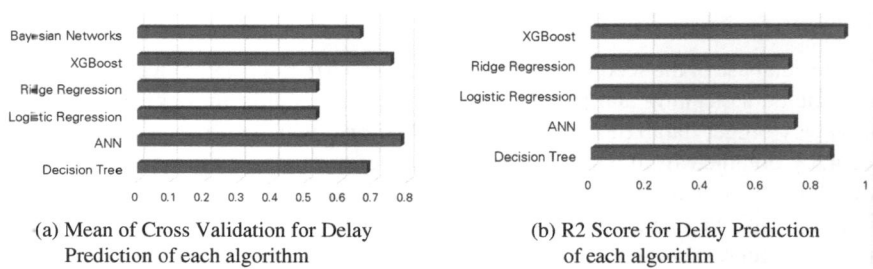

    (a) Mean of Cross Validation for Delay          (b) R2 Score for Delay Prediction
        Prediction of each algorithm               of each algorithm

**Fig. 4.9  a** Mean of cross validation for delay prediction of each algorithm. **b** R2 score for delay prediction of each algorithm

which did not correspond to reality. The outstanding performance of XGBoost can be attributed to several factors, including:

(a) Model complexity: The MLPRegressor uses a fixed architecture with a predefined number of layers and nodes. This architecture might not be optimal for the specific problem at hand, whereas the XGBoost model can better adapt to complex data patterns due to its gradient boosting framework, which combines multiple decision trees, allowing it to capture non-linear relationships more effectively.

(b) Training process: The ANN model relies on gradient-based optimization techniques, such as backpropagation, which are sensitive to the choice of hyperparameters, including learning rate, activation functions, and the number of hidden layers. In contrast, the XGBoost model uses a more robust tree-based boosting method, which is less sensitive to hyperparameter choices, and generally converges more efficiently.

(c) Interpretability and Explainability: The ANN model is often considered a "black box" due to its complex structure, making it difficult to understand and interpret its internal decision-making process. This lack of interpretability may hamper the ability to diagnose and improve the model's performance. On the other hand, the XGBoost model is built upon decision trees, which are inherently more interpretable and allow for a better understanding of the relationships between the input attributes and the target variable.

(d) Regularization: The XGBoost model incorporates regularization techniques that penalize overly complex models, reducing overfitting and improving generalization.

## 4.5 Conclusion

This chapter explored the application of probabilistic and deterministic Machine Learning (ML) models for predicting risks in construction projects, highlighting their respective advantages and disadvantages. Probabilistic models, by incorporating uncertainty in their predictions, offer a more nuanced understanding of risks, which is crucial in the inherently uncertain environment of construction projects. Being able to integrate various types of subjective and objective data, these models can provide insights into the likelihood of various risk scenarios, enabling better-informed decision-making, specifically in limited databases. On the other hand, deterministic models, with their straightforward approach, can offer clear and concise predictions, making them easier to interpret and apply. Yet, their inability to account for uncertainty and variability in risk factors can lead to oversimplified risk assessments that might overlook critical insights. Additionally, when abundant data is not available, these models fall short in providing accurate results as they do not benefit from Bayesian inference and ability to integrate subjective expert data. These findings were highlighted by previous literature in the field, and was supported by a practical comparison stage, where three different types of probabilistic and deterministic ML models were applied to two databases of different sizes.

The application of the three different models on two different case studies had two main objectives: (a) highlighting the importance of integrating multiple sources of data and judgments in accurate risk prediction and data scarcity compensation, and (b) highlighting the critical role of database size on the performance of each probabilistic and deterministic ML model.

To find answers to the first objective, a comparative analysis between the three proposed models was conducted with respect to their prediction accuracy and assigned probability to each risk. The comparison of prediction accuracy between the BN model and the deterministic ML models indicated that for small databases like the first case study, ML algorithms suffered from overfitting, while BN had an acceptable performance. Furthermore, it provided the most balanced and realistic prediction of risk probabilities due to the ability to draw conclusions based on both objective and subjective data, not undermining or overmining the probabilities.

To find answers to the second objective, all three models were applied to another case study of a much larger size. As anticipated, deterministic ML models outperformed BN when the data was abundant. Among different ML algorithms, XGBoost had the best performance due to its ability to capture both linearity and nonlinearity existing in the data.

As proved by the results, the choice of an appropriate ML algorithm depends on the nature and availability of data, the complexity of the problem to be solved, and the relationships between the input and target variables, when working with small databases with a great number of missing values, probabilistic approaches like BNs are recommended, as they can integrate various sources of judgments to compensate for the data scarcity and benefit from the Bayesian inference for a more realistic risk probability assessment. However, when working with large databases, the deterministic ML algorithms outperform the probabilistic ones due to their advanced structure and ability to capture both linearity and nonlinearity in the data. Furthermore, the existence of abundant data makes the frequentist inference of such models closer to reality.

In conclusion, while the integration of AI and ML in construction RM presents numerous advantages and holds immense potential to revolutionize the construction industry, it's essential to be cognizant of the challenges. With continued research and collaboration between experts and technologists, the construction industry can harness the full potential of digital technologies. Proper training, collaboration, and continuous evaluation can ensure that ML models serve as valuable tools in the ever-evolving landscape of construction RM. From enhanced efficiency and safety to improved quality and client satisfaction are the contributions of an intelligent and automated RM framework with the help of ML-based models. This chapter tried to delineate a portion of the numerous advantages of ML-based RM methods with supported by the findings of an extensive literature search and practical comparisons of these methods.

# References

F. Abdat et al., Extracting recurrent scenarios from narrative texts using a Bayesian network: application to serious occupational accidents with movement disturbance. Accid. Anal. Prev. **70**, 155–166 (2014). https://doi.org/10.1016/j.aap.2014.04.004

I.S. Abotaleb, I.H. El-adaway, Construction bidding markup estimation using a multistage decision theory approach. J. Constr. Eng. Manag. **143**(1), 1–18 (2017). https://doi.org/10.1061/(asce)co.1943-7862.0001204

F. Ackermann, J. Alexander, Researching complex projects: Using causal mapping to take a systems perspective. Int. J. Project Manag. **34**(6), 891–901 (2016). https://doi.org/10.1016/j.ijproman.2016.04.001

F. Afzal et al., A review of artificial intelligence based risk assessment methods for capturing complexity-risk interdependencies: cost overrun in construction projects. Int. J. Manag. Proj. Bus. **14**, 300–328 (2019). https://doi.org/10.1108/IJMPB-02-2019-0047

A. Ajayi et al., Deep learning models for health and safety risk prediction in power infrastructure projects (2019). https://doi.org/10.1111/risa.13425

T.D. Akinosho et al., Deep learning in the construction industry: a review of present status and future innovations. J. Build. Eng. **32**, 101827 (2020). https://doi.org/10.1016/j.jobe.2020.101827

Y. An et al., Determining uncertainties in AI applications in AEC sector and their corresponding mitigation strategies. Autom. Constr. **131**, 103883 (2021). https://doi.org/10.1016/j.autcon.2021.103883

H. Anysz, M. Apollo, B. Grzyl, Quantitative risk assessment in construction disputes based on machine learning tools. Symmetry **13**(5), 744 (2021). https://doi.org/10.3390/sym13050744

M. Bar-Hillel, E. Neter, How alike is it versus how likely is it: a disjunction fallacy in probability judgments. J. Pers. Soc. Psychol. **65**(6), 1119–1131 (1993). https://doi.org/10.1037/0022-3514.65.6.1119

P. Boatenga, Z. Chen, S.O. Ogunlana, An analytical network process model for risks prioritization in megaprojects. Qual. Mark. Res: Int. J. **7**(3), pp. 228–236 (2015). http://eprints.qut.edu.au/29653/.

A.J. Butler, M.K. Thomas, K.D.M. Pintar, Systematic review of expert elicitation methods as a tool for source attribution of enteric illness. Foodborne Pathog. Dis. **12**(5), 367–382 (2015). https://doi.org/10.1089/fpd.2014.1844

I.C. Cardenas, S.S.H. Al-Jibouri, J.I.M. Halman, F.A. van Tol, Modeling risk-related knowledge in tunneling projects. Risk Anal. **34**(2), 323–339 (2014)

F.R. Cecconi, A. Khodabakhshian, L. Rampini, Data-driven decision support system for building stocks energy retrofit policy. J. Build. Eng. **54**, 104633 (2022). https://doi.org/10.1016/j.jobe.2022.104633

A.P.C. Chan, F.K.W. Wong, C.K.H. Hon, T.N.Y. Choi, A Bayesian network model for reducing accident rates of electrical and mechanical (E&M) work. Int. J. Environ. Res. Public Health **15**, 2496 (2017)

D.B. Chattapadhyay, J. Putta, P. Rama Mohan Rao, Risk identification, assessments, and prediction for mega construction projects: a risk prediction paradigm based on cross analytical-machine learning model. Buildings **11**(4), 172 (2021). https://doi.org/10.3390/buildings11040172

L. Chenya, Intelligent risk management in construction projects: systematic literature review. IEEE Access **10**, 72936–72954 (2022). https://doi.org/10.1109/ACCESS.2022.3189157

J. Chou, C. Lin, Predicting disputes in public-private partnership projects: classification and ensemble models **27**, 51–60 (2013). https://doi.org/10.1061/(ASCE)CP.1943-5487.0000197

S.L. Choy, R. O'leary, K. Mengersen, Elicitation by design in ecology: using expert opinion to inform priors for Bayesian statistical models. Ecology **90**(1), 265–277 (2009). https://doi.org/10.1890/07-1886.1

J. Debnath et al., Fuzzy inference model for assessing occupational risks in construction sites. Int. J. Ind. Ergon. **55**, 114–128 (2016). https://doi.org/10.1016/j.ergon.2016.08.004

S.W.A. Dekker, Drifting into failure: complexity theory and the management of risk, in *Chaos and Complexity Theory for management: Nonlinear Dynamics*. ed. by S. Banerjee (Hershey, PA, IGI Global Business Science Reference, 2013), pp.241–253

I. Dikmen, M.T. Birgonul, S. Han, Using fuzzy risk assessment to rate cost overrun risk in international construction projects **25**, 494–505 (2007). https://doi.org/10.1016/j.ijproman.2006.12.002

M. Ebrat, R. Ghodsi, Construction project risk assessment by using adaptive-network-based fuzzy inference system: an empirical study **18**, 1213–1227 (2014). https://doi.org/10.1007/s12205-014-0139-5

M. Eybpoosh, I. Dikmen, M. Talat Birgonul, Identification of risk paths in international construction projects using structural equation modeling. J. Constr. Eng. Manag. **137**(12), 1164–1175 (2011). https://doi.org/10.1061/(asce)co.1943-7862.0000382

C. Fan et al., Deep learning-based feature engineering methods for improved building energy prediction. Appl. Energy **240**, 35–45 (2019). https://doi.org/10.1016/j.apenergy.2019.02.052

C. Fang, F. Marle, Dealing with project complexity by matrix-based propagation modelling for project risk analysis. J. Eng. Des. **24**(4), 239–256 (2013). https://doi.org/10.1080/09544828.2012.720014

D.C. Flath et al., Cluster analysis of smart metering data an implementation in practice. Eine praxisorientierte Umsetzung. WIRTSCHAFTSINFORMATIK (2012). https://doi.org/10.1007/s12599-011-0201-5

M. Gajzler, The idea of knowledge supplementation and explanation using neural networks to support decisions in construction engineering. Procedia Eng. **57**, 302–309 (2013)

A. Gelman, J.B. Carlin, H.S. Stern, D.B. Dunson, D.B.R. Aki Vehtari, *Bayesian Data Analysis*, 3rd edn. (CRC Press, Boca Raton, 2013). https://doi.org/10.1201/b16018

S. Gerassis et al., Bayesian decision tool for the analysis of occupational accidents in the construction of embankments. J. Constr. Eng. Manag. **143**(2), 1–8 (2017). https://doi.org/10.1061/(ASCE)CO.1943-7862.0001225

L. Giannakos, Y. Xenidis, Risk assessment in construction projects with the use of neural networks. in *Safety and Reliability - Safe Societies in a Changing World - Proceedings of the 28th International European Safety and Reliability Conference, ESREL 2018* (2018), pp. 1563–1570. https://doi.org/10.1201/9781351174664-197

M. Golparvar-Fard, F. Pena-Mora, S. Savarese, Automated progress monitoring using unordered daily construction photographs and IFC-based building information models. J. Comput. Civ. Eng. **29**(1), 04014025 (2015). https://doi.org/10.1061/(ASCE)CP.1943-5487.0000205

A. Gondia, A. Siam et al., Machine learning algorithms for construction projects delay risk prediction. J. Constr. Eng. Manag. **146**(1), 04019085 (2020a). https://doi.org/10.1061/(asce)co.1943-7862.0001736

A. Gondia, S.M. Asce et al., Machine learning algorithms for construction projects delay risk prediction. J. Constr. Eng. Manag. **146**(1), 1–16 (2020b). https://doi.org/10.1061/(ASCE)CO.1943-7862.0001736

A. Guzman-Urbina, A. Aoyama, E. Choi, A polynomial neural network approach for improving risk assessment and industrial safety. ICIC Express Lett. **12**(2), 97–107 (2018). https://doi.org/10.24507/icicel.12.02.97

F. Habbal et al., Applying ann to the ai utilization in forecasting planning risks in construction, in *Proceedings of the 37th International Symposium on Automation and Robotics in Construction, ISARC 2020: From Demonstration to Practical Use - To New Stage of Construction Robot*, (ISARC) (2020), pp. 1431–1437. https://doi.org/10.22260/isarc2020/0198

M. Habibi Rad, M. Mojtahedi, M.J. Ostwald, Industry 4.0, disaster risk management and infrastructure resilience: a systematic review and bibliometric analysis. Buildings **11**(9), 411 (2021)

S.H. Han et al., A web-based integrated system for international project risk management. Autom. Constr. **17**, 342–356 (2008). https://doi.org/10.1016/j.autcon.2007.05.012

F.U. Hassan, T. Le, Automated requirements identification from construction contract documents using natural language processing. J. Leg. Aff. Disput. Resolut. Eng. Constr. **12**(2), 1–12 (2020). https://doi.org/10.1061/(asce)la.1943-4170.0000379

G. Heravi, M. Asce, E. Eslamdoost, Applying artificial neural networks for measuring and predicting construction-labor productivity. J. Constr. Eng. Manag. **141**(10), 1–11 (2015). https://doi.org/10.1061/(ASCE)CO.1943-7862.0001006

O.A. Hosny, M.M.G. Elbarkouky, A. Elhakeem, Construction claims prediction and decision awareness framework using artificial neural networks and backward optimization. J. Constr. Eng. Proj. Manag. **1**(5), 11–19 (2015)

Y. Hu, D. Castro-Lacouture, Clash relevance prediction based on machine learning. J. Comput. Civ. Eng. **33**(2), 04018060 (2019). https://doi.org/10.1061/(asce)cp.1943-5487.0000810

H. Huang, H. Tserng, A study of integrating support-vector-machine (SVM) model and market-based model in predicting taiwan construction contractor default. KSCE J. Civ. Eng. **22**, 4750–4759 (2018). https://doi.org/10.1007/s12205-017-2129-x

J. Hwang, Y. Kim, A bid decision-making model in the initial bidding phase for overseas construction projects. KSCE J. Civ. Eng. **20**, 1189–1200 (2016). https://doi.org/10.1007/s12205-015-0760-y

M.S. Islam et al., A knowledge-based expert system to assess power plant project cost overrun risks. Expert Syst. Appl. **136**, 12–32 (2019). https://doi.org/10.1016/j.eswa.2019.06.030

X.H. Jin, G. Zhang, Modelling optimal risk allocation in PPP projects using artificial neural networks. Int. J. Project Manag. **29**(5), 591–603 (2011). https://doi.org/10.1016/j.ijproman.2010.07.011

K. Karakas, I. Dikmen, M.T. Birgonul, Multiagent system to simulate risk-allocation and cost-sharing processes in construction projects. J. Comput. Civ. Eng. **27**(3), 307–319 (2013). https://doi.org/10.1061/(asce)cp.1943-5487.0000218

A. Karimiazari et al., Risk assessment model selection in construction industry. Expert Syst. Appl. **38**(8), 9105–9111 (2011). https://doi.org/10.1016/j.eswa.2010.12.110

N. Khakzad, F. Khan, P. Amyotte, Quantitative risk analysis of offshore drilling operations: a Bayesian approach. Saf. Sci. **57**, 108–117 (2013)

A. Khodabakhshian, T. Puolitaival, L. Kestle, Deterministic and probabilistic risk management approaches in construction projects: a systematic literature review and comparative analysis. Buildings **13**(5), 1312 (2023). https://doi.org/10.3390/buildings13051312

E. Lamine et al., BPRIM: an integrated framework for business process management and risk management. Comput. Ind. **117**, 103199 (2020). https://doi.org/10.1016/j.compind.2020.103199

J. Lee, Y. Kim, Analysis of cost-increasing risk factors in modular construction in Korea using FMEA. KSCE J. Civ. Eng. **21**, 1999–2010 (2017). https://doi.org/10.1007/s12205-016-0194-1

S.R. Lele, K.L. Allen, On using expert opinion in ecological analyses: a frequentist approach 683–704 (2006). https://doi.org/10.1002/env.786

J. Li et al., Importance degree research of safety risk management processes of urban rail transit based on text mining method. Information (Switzerland) **9**(2), 1–17 (2018). https://doi.org/10.3390/info9020026

Z. Liu, Y. Jiao, A. Li, X. Liu, Risk assessment of urban rail transit PPP project construction based on Bayesian network. Sustainability (Switzerland) **13**(20), 11507 (2021). https://doi.org/10.3390/su132011507

M. Loosemore, E. Cheung, Implementing systems thinking to manage risk in public private partnership projects. Int. J. Project Manag. **33**(6), 1325–1334 (2015). https://doi.org/10.1016/j.ijproman.2015.02.005

L. Mkrtchyan, L. Podofillini, V.N. Dang, Bayesian belief networks for human reliability analysis: a review of applications and gaps. Reliab. Eng. Syst. Saf. **139**, 1–16 (2015). https://doi.org/10.1016/j.ress.2015.02.006

A. Mofidi et al., A probabilistic approach for economic evaluation of occupational health and safety interventions: a case study of silica exposure reduction interventions in the construction sector. BMC Public Health **20**(1), 1–12 (2020). https://doi.org/10.1186/s12889-020-8307-7

M. Mohamed, D.Q. Tran, Risk-based inspection model for hot mix asphalt pavement construction projects. J. Constr. Eng. Manag. **147**(6), 1–13 (2021). https://doi.org/10.1061/(asce)co.1943-7862.0002053

H. Nasrazadani et al., Probabilistic modeling framework for prediction of seismic retrofit cost of buildings. J. Constr. Eng. Manag. **143**(8), 04017055 (2017). https://doi.org/10.1061/(asce)co.1943-7862.0001354

N.D. Nath, A.H. Behzadan, S.G. Paal, Deep learning for site safety: real-time detection of personal protective equipment. Autom. Constr. **112**, 103085 (2020). https://doi.org/10.1016/j.autcon 2020.103085

L.D. Nguyen, D. Tran, An approach to the assessment of fall risk for building construction, in *Proceedings of the 2016 Construction Research Congress* (2016), pp. 1803–1812. https://doi.org/10.1061/9780784479827.180

A.O. Omondi, I.A. Lukandu, G. Wanyembi, Probabilistic reasoning and Markov chains as means to improve performance of tuning decisions under uncertainty. Technol. J. Artif. Intell. Data Min. **9**(1), 99–108 (2021). https://doi.org/10.22044/jadm.2020.8920.2027

Y. Pan, L. Zhang, Roles of artificial intelligence in construction engineering and management: a critical review and future trends'. Autom. Constr. **122**, 103517 (2021). https://doi.org/10.1016/j.autcon.2020.103517

J. Von Platten et al., Using machine learning to enrich building databases-methods for tailored energy retrofits. Energies **13**(10), 2574 (2020). https://doi.org/10.3390/en13102574

Project Management Institute (PMI), *A guide to the project management body of knowledge (PMBOK guide)*, 6th edn. (Project Management Institute Inc., Pennsylvania, USA, 2017)

A. Qazi et al., Project Complexity and Risk Management (ProCRiM): towards modelling project complexity driven risk paths in construction projects. Int. J. Project Manag. **34**(7), 1183–1198 (2016). https://doi.org/10.1016/j.ijproman.2016.05.008

H.M. Regan, M. Colyvan, M.A. Burgman, A taxonomy and treatment of uncertainty for ecology and conservation biology. Ecol. Appl. **12**(2), 618–628 (2002). https://doi.org/10.2307/3060967

M. Regona et al., Artificial intelligent technologies for the construction industry: how are they perceived and utilized in Australia? J. Open Innov.: Technol. Mark. Complex. **8**(1), 16 (2022). https://doi.org/10.3390/joitmc8010016

C. Sabillon et al., Audio-based Bayesian model for productivity estimation of cyclic construction activities. J. Comput. Civ. Eng. **34**(1), 1–14 (2020). https://doi.org/10.1061/(asce)cp.1943-5487.0000863

C. Samantra, S. Datta, S.S. Mahapatra, Fuzzy based risk assessment module for metropolitan construction project: an empirical study. Eng. Appl. Artif. Intell. **65**, 449–464 (2017). https://doi.org/10.1016/j.engappai.2017.04.019

A.F. Serpella et al., Risk management in construction projects: a knowledge-based approach. Procedia Soc. Behav. Sci. **119**, 653–662 (2014). https://doi.org/10.1016/j.sbspro.2014.03.073

B. Sherafat et al., Automated methods for activity recognition of construction workers and equipment: state-of-the-art review. J. Constr. Eng. Manag. **146**(6), 03120002 (2020). https://doi.org/10.1061/(asce)co.1943-7862.0001843

G. Tardioli et al., A methodology for calibration of building energy models at district scale using clustering and surrogate techniques. Energy Build. **226**, 110309 (2020). https://doi.org/10.1016/j.enbuild.2020.110309

M. Valpeters, I. Kireev, N. Ivanov, Application of machine learning methods in big data analytics at management of contracts in the construction industry, in *MATEC Web of Conferences*, vol. 170 (2018). https://doi.org/10.1051/matecconf/201817001106

F. Wang et al., Probabilistic risk assessment of tunneling-induced damage to existing properties. Expert Syst. Appl. **41**(4 Part 1), 951–961 (2014). https://doi.org/10.1016/j.eswa.2013.06.062

P. Wang et al., A Bayesian belief network predictive model for construction delay avoidance in the UK. Eng. Constr. Archit. Manag. **29**, 2011–2026 (2021). https://doi.org/10.1108/ECAM-10-2020-0873

Y.Y. Wee et al., A method for root cause analysis with a Bayesian belief network and fuzzy cognitive map. Expert Syst. Appl. **42**(1), 468–487 (2015). https://doi.org/10.1016/j.eswa.2014.06.037

B.W. Wisse et al., Relieving the elicitation burden of Bayesian belief networks, in *BMAW'08: Proceedings of the Sixth UAI Conference on Bayesian Modeling Applications Workshop*, vol. 406 (2008), pp. 10–20

D.D. Wu, S.H. Chen, D.L. Olson, Business intelligence in risk management: some recent progresses. Inf. Sci. **256**, 1–7 (2014). https://doi.org/10.1016/j.ins.2013.10.008

N. Xia et al., Towards integrating construction risk management and stake- holder management: a systematic literature review and future research agendas. Int. J. Project Manag. **36**, 701–715 (2018). https://doi.org/10.1016/j.ijproman.2018.03.006

Z. Yang, S. Bonsall, J. Wang, Fuzzy rule-based Bayesian reasoning approach for prioritization of failures in FMEA. IEEE Trans. Reliab. **57**(3), 517–528 (2008). https://doi.org/10.1109/TR.2008.928208

Z.M. Yaseen et al., Prediction of risk delay in construction projects using a hybrid artificial intelligence model. Sustainability **12**(4), 1–14 (2020). https://doi.org/10.3390/su12041514

A.E. Yildiz et al., A knowledge-based risk mapping tool for cost estimation of international construction projects. Autom. Constr. **43**, 144–155 (2014). https://doi.org/10.1016/j.autcon.2014.03.010

T. Yu et al., Evaluating different stakeholder impacts on the occurrence of quality defects in offsite construction projects: a Bayesian-network-based model. J. Clean. Prod. **241**, 118390 (2019). https://doi.org/10.1016/j.jclepro.2019.118390

F. Yucelgazi, I. Yitmen, An ANP model for risk assessment in large-scale transport. Arab. J. Sci. Eng. **44**, 4257–4275 (2020). https://doi.org/10.1007/s13369-018-3314-z

L. Zhang, M.J. Skibniewski et al., A probabilistic approach for safety risk analysis in metro construction AND. Saf. Sci. **63**, 8–17 (2014a). https://doi.org/10.1016/j.ssci.2013.10.016

L. Zhang, X. Wu et al., Bayesian-network-based safety risk analysis in construction projects. Reliab. Eng. Syst. Saf. **131**, 29–39 (2014b). https://doi.org/10.1016/j.ress.2014.06.006

L. Zhang et al., Towards a fuzzy Bayesian network based approach for safety risk analysis of tunnel-induced pipeline damage. Risk Anal. **36**(2), 278–301 (2016). https://doi.org/10.1111/risa.12448

# Chapter 5
# Computer Vision for Asset and Facility Management

The EN ISO 41011:2017 (CEN 2017) defines the Facility Management (FM) discipline as an:

> organizational function which integrates people, place, and process within the built environment to improve the quality of life of people and the productivity of the core business

Typically, FM's role is to manage the facilities and services required for the company's operations and contribute to the design and building phases. This procedure ensures that an appropriate level of service is maintained to fulfil the company's needs, while also guaranteeing that the work environment is improved while expenses are controlled through an integrated approach. As can be observed, the definition is heavily geared toward managing special purpose properties, which must deliver good performance as a result.

Besides, the process of controlling and measuring the performance of a product or service is referred to as Condition Inspection and Monitoring. It must be carried out following a set of guidelines. The ISO 15686-3, 7, 10 standard contains general rules for this function (ISO 2002, 2010, 2017). Condition Inspection and Monitoring refers to a set of procedures that should be followed in order to assess a product's or service's capacity to perform as intended under real-world situations. This essential digital Asset Management (AM) area is included in the audit process when managing an organization, and it can be engaged at various stages of the asset's life cycle. In this context, asset performance evaluation and reporting play an important role in fostering a comprehensive understanding of physical objects and the prevention of potential defects caused by unforeseeable events. Determining the purpose and the amount of detail to carry out the activities is critical in this core area. This enables the necessary data to be collected and transmitted in the Key Performance Indicators (KPIs) used to measure the appropriate asset performance and, ultimately,

By Luca Rampini.

© The Author(s), under exclusive license to Springer Nature Switzerland AG 2024
F. Re Cecconi et al., *Building Tomorrow: Unleashing the Potential of Artificial Intelligence in Construction*, PoliMI SpringerBriefs,
https://doi.org/10.1007/978-3-031-77197-2_5

make educated decisions. This area is critical for gathering precise information, understanding digital and physical assets, and tracking their performance throughout their life cycle. It also benefits other key digital AM areas such as FM, LCC, and Risk Management since it allows responsible parties to receive relevant data for conducting evaluations and assessments at various decision-making levels.

Operative functions are utilized to meet short-term goals and daily operations. Vast amounts of disorganized data can be generated from operational activities, and BIM technology is currently being used in building Operations & Maintenance (O&M) to manage this data. BIM provides a parametric and detailed model with the related building components and integrated model views that enable constant synchronization of any changes within a unified information repository. Furthermore, BIM allows for better information sharing from the design and construction phases to the O&M phase and the storage of large amounts of data created during the O&M phase (Peng et al. 2017).

Because of their high potential for performance improvement and considerable environmental effects, existing buildings are increasingly prioritized in the AECO (Architecture, Engineering, Construction, and Operations) industry (Dinc et al. 2020). However, AECO applications on existing structures often necessitate 3D models that accurately depict the as-is conditions. Manual modeling based on building documentation (i.e., drawings, specifications, schedules) and audits is a labor-intensive and challenging process susceptible to imprecision and error (Brilakis et al. 2010). Therefore, numerous studies have looked into how to replicate the semantics and geometry of interior settings in recent years. These studies fall into three categories: primitive-based, geometry-based, and deep learning-based techniques (Mahmoud et al. 2024).

Primitive-based approaches rely on shape descriptors and primitives to decompose objects into simpler shapes, which are then combined to reconstruct the original object (Nan et al. 2012). This method is particularly effective for objects with simple geometric structures but can lead to inaccuracies with more complex forms. Geometry-based techniques aim to recover the complete 3D geometry of an environment by modeling polygonal structural elements such as walls, floors, and ceilings (Wang et al. 2015). These methods often incorporate geometric priors like point normal, adjacency, and spatial distance, but they can struggle with the automation and handling of irregular indoor objects.

In contrast, deep learning-based techniques excel in 3D model reconstruction due to their ability to autonomously learn complex features and better understand semantics. They are highly flexible across different settings and can efficiently manage extensive point cloud data. Significant advancements have been made in point-cloud segmentation, with direct methodologies like point-based and graph convolution network-based methods showing particular promise. For instance, methods such as Point-Net (Qi et al. 2017a, b) and its successors (Point-Net++, Point-CNN) have set new standards in this domain (Qi et al. 2017a, b; Li et al. 2018). Additionally, graph neural networks enhance performance by considering both points and edges in their analyses.

In this context, Computer Vision (CV) approaches, particularly 3D reconstruction algorithms, have potential in the data acquisition and modeling of building or component geometry. CV is a broad term that refers to various techniques for extracting and processing visual data from images and videos to make inferences. Some of these techniques that are important to construction management tasks are: (1) 3D scene reconstruction, (2) Image and object classification, (3) Object recognition, (4) Object tracking, (5) Segmentation, and (6) Action recognition (Paneru and Jeelani 2021). Nonetheless, training reliable CV models for O&M tasks requires a lot of visual data that is currently unavailable in the field. The lack of data for some AECO applications is one of the most perceived problems that emerged from the AIRI analysis. This chapter proposes a possible solution based on the combined use of synthetically generated images and generative AI that can be adopted for many AECO tasks.

## 5.1  Synthetic Images for Semantic Understanding in Facility Management

The Operations and Maintenance (O&M) stage represents the most extended phase in the life cycle in the Architecture, Engineering, Construction, and Operations (AECO) sector (Akcamete et al. 2019). During this phase, many stakeholders handle processes and procedures, often appearing and leaving at different times, causing a loss or a distortion of asset information. According to the National Institute of Standards and Technology (NIST), approximately 57.8% of the projected $15.8 billion annual costs are incurred by owners and operators during the operational phase (Gallaher et al. 2004). These expenses are brought on by ineffective business process management, redundant facility management systems, lost productivity, rework costs, and other problems. Nowadays, a solution for better information management is represented by Digital Twins (DTs)—an updated and accurate digital replica of a physical asset that represents the asset's as-is condition (Brilakis et al. 2019). However, it is necessary to detect objects and their geometric relationships within the asset to generate DTs. Most research is focused on recognizing large architectural components such as columns, ceilings, and walls rather than detecting secondary building components like Heating, Ventilation, and Air Conditioning (HVAC) elements, which are a crucial part of effective Facility Management (FM). However, the production and the update of precise and reliable DTs that reach FM information level present different challenges:

– Compared to the design and construction phases, the operation stage is dynamic, with many changes in uses, tools, and pieces of furniture of the various parts that constitute the asset.
– DT is characterized by many objects and systems that are usually smaller than structural elements, making their representability and updatability difficult.

– Compared to structural components, FM-related assets have a wider range of variance within classes, necessitating learning additional feature patterns. For instance, radiators will have slightly varied markings, valve designs, and other features.

The significant manual effort required to create an enriched DT is prohibitively expensive compared to the resulting model's perceived value. For these reasons, there is a high demand for greater automation in creating an information-rich DT. Using image processing and machine learning techniques, researchers have recently studied methods for extracting features such as colour, texture, and shape that can distinguish target components from other objects. In this context, Deep learning models have been applied for contextual awareness of scenes as computers have advanced with the introduction of faster GPUs. Such applications necessitate the collection of a large amount of labelled image data. In this context, large-scale publicly available datasets like the Scene Understanding (SUN) database (Song et al. 2015), Common Objects in Context (COCO) dataset (Lin et al. 2014), KITTI Cityscapes dataset (Geiger et al. 2012), and (Caesar et al. 2020) have been generated. However, data on FM-related scenes are scarce. Thus, there is a vital requirement for large-scale annotated image data regarding the asset's operations components.

The variety of types, shapes, and materials of FM-related objects complicates the implementation of a recognition model that performs well. While preparing data for asset components scene understanding, two challenges arise: the first is that image labelling is done by hand, which is time-consuming and expensive. Image labelling for object recognition is the task of using a polygon to mark the area containing the object and specify the class of the object. A dozen clicks on a single object are required to mark a polygon. The second problem is that domain knowledge is necessary to label the FM items. Identifying the area and class of objects in a scene photograph requires expertise. For example, identifying a colour code that indicates the contents of a pipe in a plumbing system requires competence. This knowledge may require additional training for the labeller. These two challenges can be addressed using existing 3D object drawings that are often used inside BIM models. Indeed, in the last few years, many vendors provided detailed and accurate digital representations of FM components. Those virtual models are collected in several open datasets that can be used as a source for generating FM-BIM models. Hence, the labelling operation can be performed in a BIM environment using a virtual camera. However, variances in colour and texture between BIM and real-world images—usually photographs—cause differences in spatial elements that serve as training requirements. As a result, for BIM images to be used as training data for photograph analysis, they must first be translated into a photographic style.

Recently, it is increasingly common to use open-source graphics engines, such as Blender, Unity, or Unreal, in computer vision and machine learning to generate synthetic data. Synthetic data is information that is artificially manufactured rather than generated by real-world events, which has several significant benefits, including: (1) the option to automatically generate labelled data; and (2) the possibility to respond to different environmental factors, such as various lighting situations, seasons, and day-night cycles. To date, no synthetic data pipeline has been developed in the context of enriching FM-BIM models. Therefore, this study addresses the following research questions:

- What pipeline can be used to generate FM-related synthetic data?
- Are FM synthetic data valuable to increase the accuracy of FM components' object detection?

## 5.2 Background

A framework based on synthetic data is proposed in the following case study in order to enable and facilitate the detection of small secondary objects and to fill DT with adequate information for FM operations. Most previous studies (Wang et al. 2017; Hou et al. 2020; Hong et al. 2021; Pan et al. 2021) concentrated on detecting structural objects such as floors, ceilings, and walls, while only a few studies paid attention to small objects that are components of assets sub-systems, such as firesafety or energy systems. Compared to structural components, FM-related objects are typically smaller than structural parts. Therefore, more training data are required to use similar OD techniques to find such small components. Consequently, synthetic data, if effective, can considerably aid in training object detection models with a suitably diversified and balanced dataset without relying on manual collection and annotation of large datasets. The object detection applications in the AECO domain and the introduction and use of synthetic data in current research are reviewed in the following paragraphs.

*Object detection models*

The objective of generic object detection is to locate and recognize objects present in a single image, then label them with rectangular bounding boxes to denote their certainty of existence. There are two strategies for building the frameworks of generic object identification algorithms: the first strategy generates region proposals first, then categorizes each proposal using the typical object detection pipeline. The second strategy views the identification of objects as a regression or classification problem, using a unified framework to arrive quickly at conclusions (categories and locations) (Zhao et al. 2018). For this study, we decided to deploy the classification-based method because it can process images in real-time, which is beneficial for the surveying activities conducted in FM's scope (Fig. 5.1). The presented case study deploys the You Only Look Once (YOLO) model, developed by Redmon et al. (2016). YOLO employs the highest feature map to forecast confidences and bounding boxes for several categories. Since its debut, it has continuously improved thanks to numerous cutting-edge techniques such as batch normalization, anchor boxes, multi-scale training, etc. We specifically utilized the fourth iteration made available by Bochkovskiy et al. (2020). In a nutshell, the input image is divided into a $S \times S$ grid by YOLO, and each grid cell is responsible for guessing the object

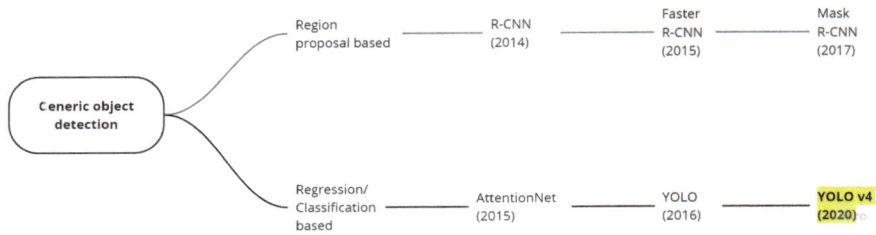

**Fig. 5.1** The two object detection frameworks are region proposal and regression/classification

centred in that grid cell. Each grid cell forecasts B bounding boxes and their associated confidence scores. Despite the YOLO v4 model's availability, it is impossible to fully utilize its weights because they were trained to recognize categories that do not apply to our application domain. However, by using a transfer learning technique, we can employ a significant amount of the weights that have already been trained. Transfer learning is the practice of using previously acquired knowledge to address brand-new, related problems. Since the model has already "seen" and "learned" from numerous photographs, we can benefit from a pre-trained model that was previously trained on thousands of images.

*Detection of secondary objects in buildings*

While most research is used to help robots recognize certain objects in the environment and carry out a specific activity, more needs to be done in the AECO field. Adán et al. (2018) proposed a method for finding items in a colored point cloud, such as switches, ducts, and signs. Depending on whether the objects have geometric or color discontinuities in the wall area, potential zones of interest are calculated using detailed photos and color images of the wall plane. The region of interest is then compared to a predefined depth model database and a predefined color model database containing object classes from the scene. Moreover, the DenseNet model (Huang et al. 2017) for feature extraction was implemented by Wei and Akinci (2019) on various publicly accessible datasets. The authors put out a framework for semantic segmentation-based image-based localization and understanding. Although only a portion of the object is required to link it to its DT, the conclusion noted that object detection models could enhance the pipeline previously discussed. As a result, a coarser bounding box may be sufficient to permit the linkage. Finally, (Pan et al. 2022) suggested a pipeline to improve Geometrical DT by utilizing photos and useful text data. Although the accuracy increases for objects that vary significantly in different surroundings (such as lights and sockets), the deployed datasets needed to be sufficient to train the model despite achieving good results in several object categories, such as fire extinguishers and smoke alarms. Therefore, the proposed method required more information to be effective, particularly to cover other objects that were ignored in the study but are still crucial to DT (e.g., bookshelves, desks, etc.). Our study aims to fill this gap by proposing artificially generated synthetic data.

*Synthetic data*

In the computer vision and machine learning sectors, it has become common practice to employ fabricated data to enhance the performance of a taught model (Rampini and Cecconi 2022). Due to the large amount of data needed to train a deep learning model, many academics have suggested using synthetic data to supplement existing datasets and provide training data for new applications. Bridging the reality gap with real-world data is difficult when using synthetic data. The main selling advantage of artificially generated data, the ability to generate a substantial amount of already labelled data effectively for free, is negated by the costs in terms of time and the computer power required to generate a sufficient amount of photorealistic data (Tremblay et al. 2018). Therefore, recent approaches focused on creating diverse scenarios by changing objects' 3D models (Peng et al. 2015) and backgrounds (Salehi and Burgueño 2018).

There are few studies on synthetic data in the AECO field, and none of it is concerned with FM-related objects. A process for autonomously producing tagged and high-quality synthetic data was suggested by Hong et al. (2021). The study involved three basic steps and focused on the structural components of buildings and bridges: (i) utilize CycleGAN to transform BIM images into real-world photos; (ii) automatically label them using the spatial data in the BIM to construct various synthetic datasets; and (iii) combining the final synthetic dataset created by splicing the selected synthetic datasets. Other studies created artificial datasets for a variety of uses: (Neuhausen et al. 2020) enhanced the performance of a YOLOv3 detector in tracking and monitoring workers' movements on the constructions site by adding around 600 synthetically generated in 8 on-site construction scenes; (Sutjaritvorakul et al. 2020) deployed a fully synthetic dataset to identify workers from a load-view crane camera, improving the safety of the crane's operation. Additionally, recent studies have demonstrated the potential of artificially created indoor scenes, where images with various furniture and lighting were given with highly photorealistic footage (Li et al. 2019; Roberts et al. 2020). Finally, Wei and Akinci (n.d.) proposed a pipeline for generating synthetic data using 4D-BIM for scene understanding. In particular, the fourth dimension of BIM is leveraged to deal with the dynamic environment that is usually characterized on-site construction field. However, to date, no workflows focus on synthetic images representing FM secondary objects; hence, this study aims to fill this gap.

**A pipeline for producing synthetic images**

This research is part of a broader schema that automatically enriches DT models with FM-related information (Fig. 5.2).

The pipeline proposed in this study covers most of the Dataset creation and Model definition phases. The Model use, which includes the geometric relationships among the objects, is left outside the scope of this study, and it is partially addressed in Pan et al. (2022). The workflow contains two primary steps: (i) the procedural generation of synthetic images through a graphic engine (Blender) and 3D CAD models, and (ii) the training of an object recognition model made by a combination of real

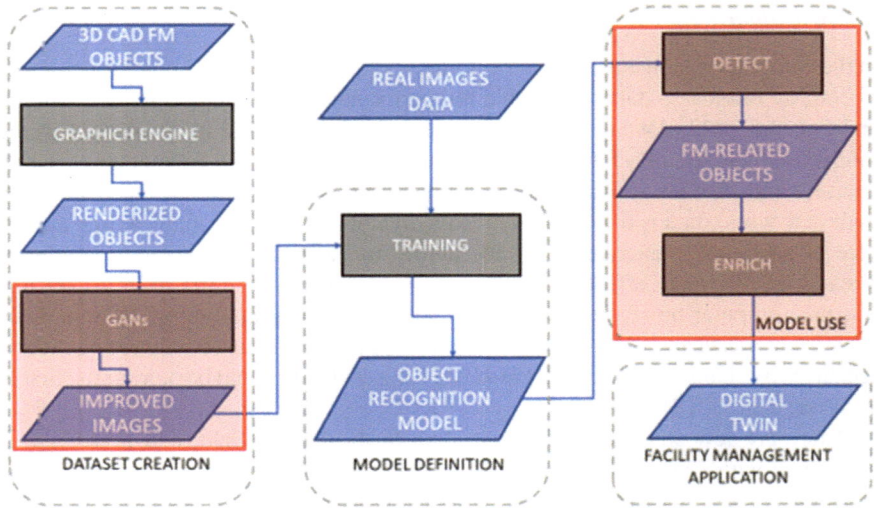

**Fig. 5.2**   The overall research schema (the parts colored in red are not covered)

and synthetic images. In the following subparagraphs, the two steps are furtherly explained. Over the past decade, DL models have achieved impressive performance and evolved from simple classification tasks to more complex topics such as object detection and/or segmentation. For most applications, rather than increasing the model's complexity, the focus has shifted to providing sufficient data to train the model, especially for computer vision tasks. The availability of an extensive, well-balanced, and precisely labelled dataset is an ideal situation that is rarely verified in real-world applications. Often, datasets are scarce or require much time and annotation effort. For example, the ImageNet dataset, one of the most important in the CV field, required almost two years to label around 11 million images with crowd-sourcing methods (Russakovsky et al. 2015). As a result, many researchers are trying to avoid the challenges of collecting and annotating data in the real world by creating virtual environments that generate training examples in a controllable and customizable manner. These artificially generated data are known as synthetic datasets and present several advantages over real-world based data:

- Produce balanced datasets
- Cover a wide range of lighting conditions
- Generate automatically labelled images
- Produce more semantic representations such as depth maps, segmentation maps, and so on
- It is easier to comply with privacy regulations such as the EU General Data Protection Regulation (GDPR) (Voigt and Bussche 2020) or the California Consumer Privacy Act (CCPA) (Bukaty 2019). Especially in a little-shared industry such as

construction, it is challenging to collect publicly available indoor images from different buildings.

Figure 5.3 shows the workflow's difference between a real-world, manually annotated dataset and a synthetically generated one. Generally, the input data for generating synthetic images is an environment built with 3D assets. In the AECO industry, thanks to the growing adoption of CAD first and BIM lately, a considerable amount of 3D drawings are widely present in the market. Moreover, most of these models are available in opensource datasets formed by models freely provided by vendors. Therefore, it is possible to leverage the existing 3D BIM models and use them as a backbone for generating AECO-related synthetic datasets. In this study, we used the BIMobject platform (BIMobject 2015), which contains different object categories (with different formats) that covers several building systems structural, electric, heating, and so on). From BIMobject, it was possible to download the 3D models and import them into a graphic engine. A graphic engine uses computer time rather than human time to generate examples. It has complete information about the scenes it renders, allowing it to save time and money on human annotations and reviews. A graphic engine also allows for the generation of rare examples, allowing control over the training dataset's distribution.

For this study, we used Blender—a free and open-source 3D graphic engine that supports three-dimensional object modeling, simulation, and rendering (blender.org 2015). The engine was chosen among the others for its embedded Python API, which allows scripts that facilitate the process of iterating through light conditions, camera poses, and textures, as well as the process of generating annotations.

Despite the abovementioned advantages of synthetic data, training and running an object detection model that relies entirely on artificially generated data cannot guarantee good performances. Even if the real-world data are limited to 10% of the entire dataset, the benefits in terms of precision and recall are well documented (Nowruzi et al. 2019). Although there are no existing datasets for object detection focusing on FM-related objects, inside the largest annotated image dataset—Google Open Images V6—we can use the "power plugs and sockets" category to identify those elements. The dataset was released in 2020 and comprised 1.9 million images for 16 million manually annotated bounding boxes for 600 object categories. For the "power plugs and sockets" category, 112 images are available for training, totalling 198 bounding boxes. On the other hand, the pipeline proposed to create synthetic images tailored explicitly for FM applications is shown in Fig. 5.4.

Eevee and Cycles are the two main render engines accessible in Blender. In general, Eevee was designed for real-time rendering, whereas Cycles was designed for realism. As a result, unless dedicated graphics cards (GPUs) are employed, Cycles renders images substantially slower than Eevee. Currently, a user can build all accessible forms of ground truth maps when using Cycles, unlike Eevee, which cannot generate segmentation masks or optical flow ground truths. (Mayer et al. 2018) presented a thorough examination of several synthetic and real-world datasets for training neural networks for optical flow and disparity estimation applications. The

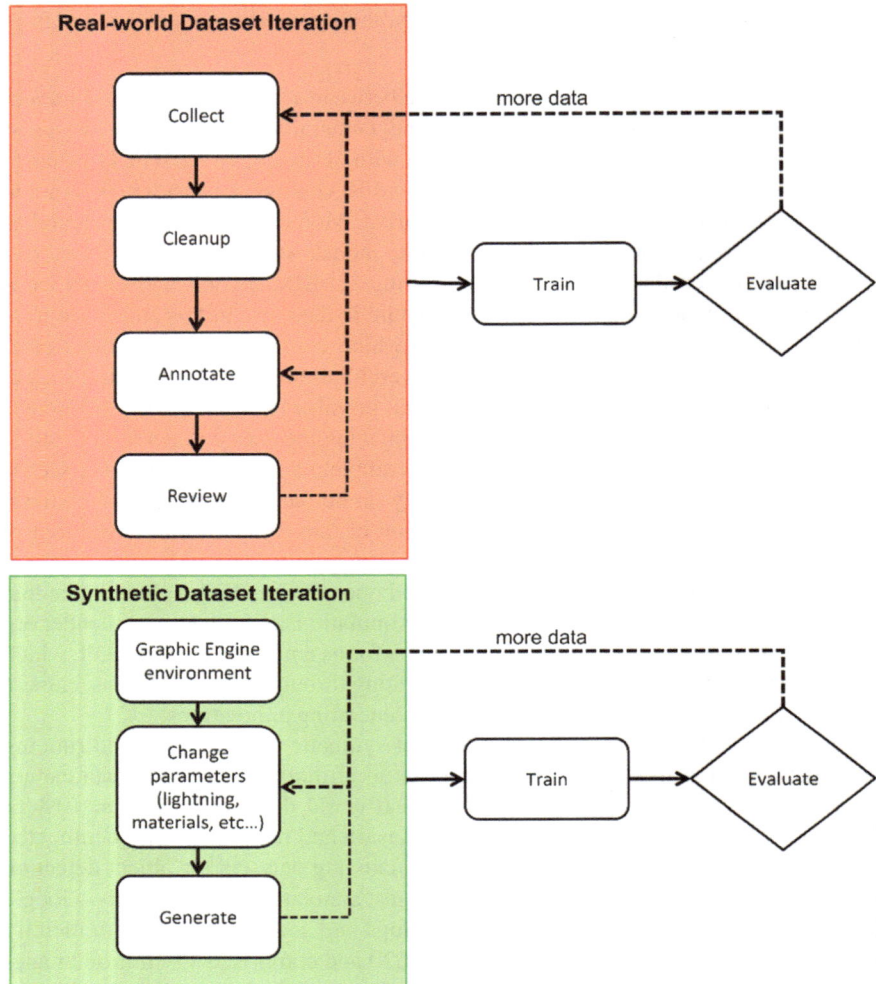

**Fig. 5.3** The dataset iteration process is compared using real-world and synthetic datasets (derived from Borkman et al., 2021). Top: A real-world dataset necessitates collecting, cleanup, annotating, and reviewing. These steps necessitate costly and time-consuming human labor. Bottom: Creating a synthetic dataset requires creating an environment with 3D assets, adjusting randomization parameters, and running the environment to generate new data. The datasets include accurate annotations and are validated automatically, eliminating the most time-consuming steps

authors highlight two significant findings. The first point is the significance of diversity in the training set. Second, they demonstrate that the photo-realism of the dataset is not significantly influencing the model's strong performance. Therefore, in order to introduce a pipeline as accessible and flexible as possible, we used the Eevee render engine to build the synthetic dataset. The output of the generation process is a series

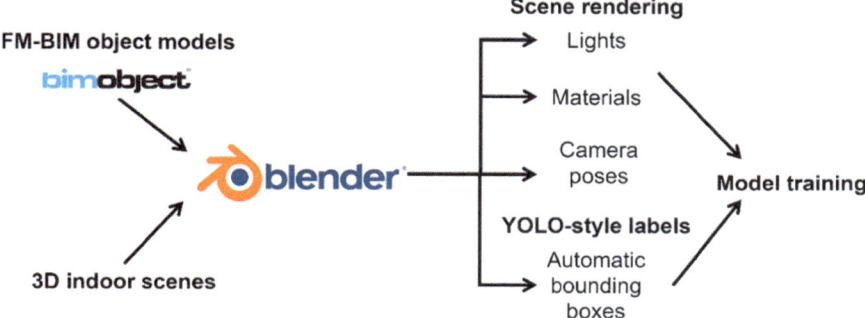

**Fig. 5.4**  Proposed pipeline for creating FM-related synthetic images

of RGB synthetic images and associated text files that contain the coordinates of the bounding boxes surrounding the sockets. The text files are produced in the format compatible with You Only Look Once (YOLO), where each object's bounding box inside the picture is reported in the form: (i) object class, (ii) x coordinate of the bounding box centre, (iii) y coordinate of the bounding box centre, (iv) bounding box width, (v) bounding box height. Consequently, to test the accuracy of using a combination of real and synthetic images, we created 100 images (like the ones in Fig. 5.5) by rendering and iterating through 10 different types of sockets, 10 different settings (kitchen, bathroom, bedroom, and living room) with different camera poses and light conditions.

The model implemented for this study, which is the fourth version of the YOLO architecture (Bochkovskiy et al. 2020), has been chosen for two main reasons:

1. It can detect objects almost in real-time using videos and photos: this aspect is significant considering how the surveys are conducted in our field, where the use of 360 cameras and drones is growing dramatically.
2. Despite the newer version of YOLO (up to v7), the fourth version is more robust and has been used and proven in several applications. However, changing the version of the model in the future should not change too much the pipeline steps proposed in this study.

To test the effectiveness of introducing synthetic data, we must define the metrics we use to evaluate the model. In object detection tasks, the predictions are made using a bounding box and a class label. The overlap between the predicted and the ground truth bounding boxes determines the accuracy of the prediction, which is commonly called Intersection over Union (IoU) (Fig. 5.6).

A threshold for the IoU value needs to be defined in order to calculate the Precision and Recall of the object detection model. For instance, if the IoU threshold is 0.5, and the IoU value for a prediction is 0.8, then it is considered a True Positive (TP). On the other hand, if the IoU is 0.4, the prediction is classified as False Positive (FP). Therefore, the prediction's precision and recall differ by setting different IoU thresholds. Usually, the precision-recall curve is adopted to represent the tradeoff

**Fig. 5.5** Examples of synthetically generated data

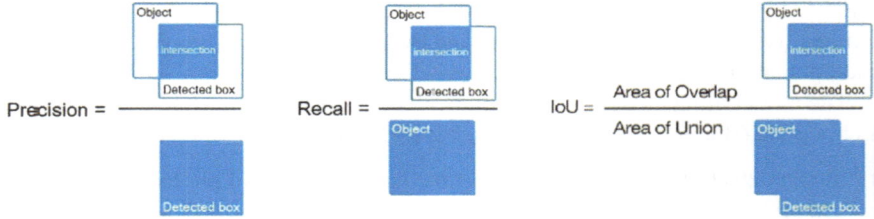

**Fig. 5.6** Calculation process of precision, recall and IoU

between precision and recall for different thresholds. A high area under the curve indicates both strong recall and high precision, with high precision corresponding to a low false positive rate and high recall corresponding to a low false negative rate. The Average Precision (AP) is defined by the area under the precision-recall curve. For multi-classes detection tasks, the most common metric is the mean Average Precision (mAP) score, defined below:

$$mAP = \frac{1}{N} \sum_{i=1}^{N} AP_i$$

**Table 5.1** Datasets used to train the object detection model

| Dataset | Real images | Synthetic images |
|---------|-------------|------------------|
| 1 | 112 | 0 |
| 2 | 66 | 66 |
| 3 | 112 | 100 |
| 4 | 112 | 888 |

where N is the number of classes. Since we are detecting one class, the AP is equivalent to the mAP. In this study, we evaluated the mAP for three different cases, summarized in Table 5.1.

The first dataset includes only real-world images and is used as the benchmark. The second dataset is also composed of 66 real images and 66 synthetic images, making the total number of images the same as Dataset 1. This dataset helps us understand how the introduction of synthetic images affects the model's performance when the dataset size remains unchanged. Although it's an unlikely practical scenario (replacing real images entirely with synthetic ones), it provides insights into the impact of synthetic images. Datasets 3 and 4 are mixed datasets that include both real and synthetic images. In Dataset 3, there are 112 real images and 100 synthetic images, meaning the synthetic images make up approximately 50% of the total, but now they are in addition to the original size of Dataset 1. In Dataset 4, there are 112 real images and 888 synthetic images, so synthetic images comprise about 90% of the total. This helps us analyze the model's performance when synthetic images vastly outnumber real ones. This range of datasets allows us to thoroughly investigate the effects of synthetic images on model performance under various conditions.

## 5.3   Object Detection with Synthetic Images

In this section, we evaluate the performances of the YOLO v4 object detection model with the four datasets. We use the mAP, a common evaluation metric for object detection tasks. The comparison clarifies if the proposed pipeline can be used for FM-related tasks, perhaps extending the methodology to other components such as fire extinguishers, furniture, or heating equipment. Finally, we discuss the results and the insights that can be derived from them.

The training of the object detection model has been performed using the NVIDIA Tesla P100 graphic card with 16 GB of vRAM. The model requires as input 416 × 416 RGB images, which are divided into batches of 32 images. The number of epochs for training is set to 2000 after trying different values (more epochs were causing overfitting, were less a drop in performance), and the learning rate was set to 0.001.

Figure 5.7 shows the mAP performance of the validation set during the training. The validation set is formed by 20% of the real images training set, except for Dataset 4, where the validation set is 40% of the real images (i.e., 5% of the overall training dataset). We decided to use only real images for the validation set since also the test dataset included them. In this way, the performances on the validation set are comparable to those on the test set. In all datasets, the models are increasing the mAP value until it reaches a stable value with the epoch increase, meaning that the model has converged.

Moreover, Table 5.2 shows the mAP performances on the test dataset. Noteworthy, the mAPs on the test and validation datasets are similar, assessing the robustness of model performance.

The worst performance is in the case of the smallest dataset (Dataset 2), composed equally of real and synthetic images. Considering that a dataset composed of only

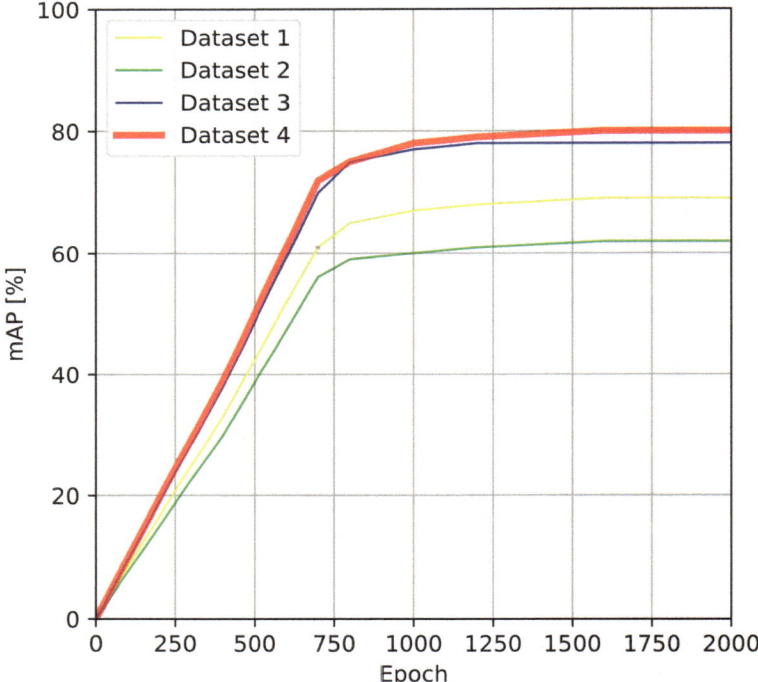

**Fig. 5.7** MAP performance on the validation set for each dataset. The model is learning most of its weights in the first 1000 epochs and reaches a stable value in the last epochs

**Table 5.2** MAP performances of the four datasets

| Dataset | mAP |
|---------|-----|
| 1 | 69% |
| 2 | 61% |
| 3 | 77% |
| 4 | 79% |

real images (Dataset 1) performs better, it follows that, for the same size, datasets formed of only real images perform better. However, it should be remembered that the primary purpose of synthetic images is to enable model training when sufficient real images are unavailable. Therefore, it is more interesting the results obtained with Datasets 3 and 4 compared to Dataset 1. In this case, there is an increase in performance with both Dataset 3 and Dataset 4.

It should be noted that there is no significant difference between Dataset 3 (composed of 50% synthetic images) and Dataset 4 (composed of 90% synthetic images). Therefore, we can infer that the benefits of adding synthetic images are most significant when the number of images added is comparable with the initial dataset. In contrast, the benefits are significantly less when adding more images than the original dataset.

The results are also affected by the distance between the cameras and the target object. Since sockets and power plugs are small objects, the training images were taken not too far from the objects (around 50 cm). Therefore, introducing pictures of an entire room may not trigger the detection of the targeted objects. However, the speed of the YOLO network (close to real-time) allows for adjusting the distance quickly by giving immediate feedback to the user (especially with video cameras).

## 5.4  Discussion

The previous paragraph shows that the introduced pipeline can be deployed to enhance the object detection model's performance by leveraging synthetically generated data. The experiments, aiming at recognizing sockets and power plugs, gave numerous insights and reflection points.

First, for the same number of images, the dataset composed of only real images performed better than the mixed dataset (real and synthetic), meaning that the virtually generated images are a powerful tool for enhancing and integrating existing real image-based datasets rather than replacing them. This is evident from Dataset 3 and Dataset 4, where the mAP score increased by a factor of 10% compared to Dataset 1. Therefore, synthetic images are a viable solution to address the challenges that FM-related object detection models face for effective training. Moreover, this performance gap might be reduced by addressing the sim-to-real gap with other AI techniques, such as Generative Adversarial Networks (GANs). Second, the possibility to easily create annotated synthetic data allowed us to embrace the FM object's variability. For instance, in some scenarios like the ones in Fig. 5.8, the model trained on Dataset 1 could not correctly predict sockets and power plugs because neither black nor horizontally oriented sockets were present in the training dataset. Solving these problems using only real-world images would have required collecting and annotating images that include the cases above, spending considerable time and resources. On the other hand, synthetic images were sufficient to change color or orientation to the 3D BIM models already used. Moreover, creating images capable of recognizing these types of scenarios also took very little time. This means that synthetic data can

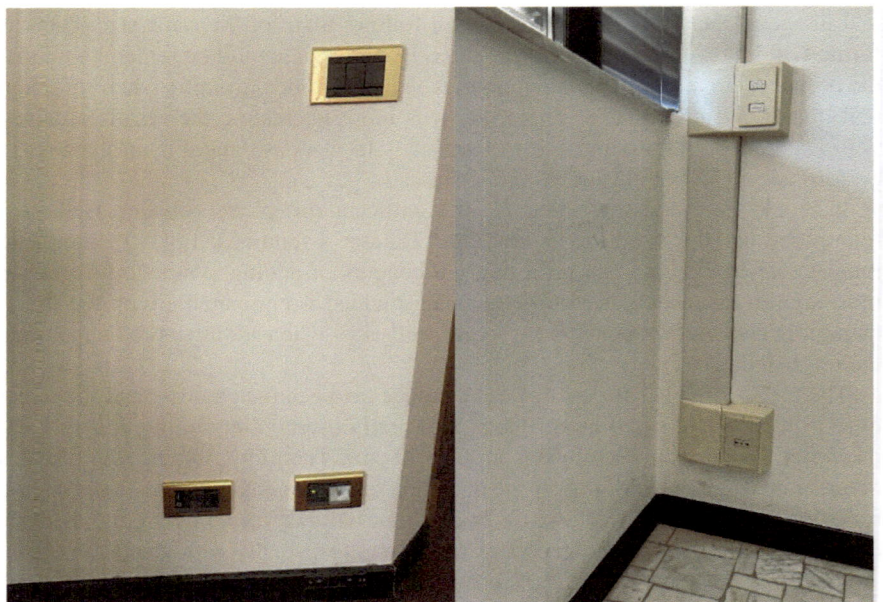

**Fig. 5.8** Examples of socket types not included in dataset 1 and, therefore, not recognized by the model. By adding synthetic images based on 3D drawings similar to the power plugs in the picture, the model can recognize them

easily solve problems related to the variability of FM objects, perhaps iterating the training process once problems are found through feedback and testing.

Finally, it is worth mentioning that the performance of Dataset 3 and 4 are comparable, meaning that many synthetic images do not guarantee a comparable increase in performance. Consequently, it is hard to establish a minimum required number of images to achieve the desired results because they depend on different aspects like object size, typologies, etc. Therefore, defining the correct number of synthetic images to include will be established by an iterative process (somewhat like what happens when defining the depth and density of layers of Neural Networks).

In conclusion, synthetic images facilitate the introduction of more data quickly and enable iteration throughout the process. If the performance is not good enough, more images can be created to train the model again without conducting another campaign for data collection and annotation.

## 5.5   Conclusion

The documentation of as-is conditions has increasingly been done by capturing videos and images. However, to automate the process of extracting information from such data, it is necessary to train deep learning models that re-quire producing a large amount of labelled data, which is a costly and timely demanding task. In this study, we introduced a pipeline tailored for the AECO industry for generating FM-related synthetic images to overcome difficulties with classifying ground truth visual FM data. Although other methodologies have been proposed for different industries, the proposed method takes advantage of the existing 3D BIM object models that are freely accessible to create a training dataset that encompasses the widest possible collection of FM-related objects. Moreover, using a graphic engine allows the production of more realistic images by deploying advanced rendering tools that help to close the sim to real gap. The method- ology has been tested to recognize sockets and power plugs to answer the first research question. However, using only open-source and freely available sources and products, the proposed method can produce the desired amount of virtually generated data of any objects modelled in a BIM environment. The created dataset has been used to train a YOLO object detection model, and its performances are compared with those obtained using real training data. As an answer to the second re-search question, the experiment findings demonstrated that the suggested strategy outperforms models trained on only real-world photos by covering a broader object's variability and increasing prediction robustness.

## References

A. Adán et al., Scan-to-BIM for "secondary" building components. Adv. Eng. Inf. Elsevier **37**, 119–138 (2018). https://doi.org/10.1016/J.AEI.2018.05.001

A. Akcamete, B. Akinci, J.H. Garrett, Potential utilization of building information models for planning maintenance activities, in *EG-ICE 2010 - 17th International Workshop on Intelligent Computing in Engineering* (2019)

BIMobject, BIM Object (2015), https://www.bimobject.com/en. Accessed 8 Aug 2022

blender.org (2015) blender.org, Blender.Org, https://www.blender.org/. Accessed 8 Aug 2022

A. Bochkovskiy, C.-Y. Wang, H.-Y.M. Liao, YOLOv4: optimal speed and accuracy of object detection (2020)

S. Borkman, A. Crespi, S. Dhakad, S. Ganguly, J. Hogins, Y.-C. Jhang, M. Kamalzadeh, B. Li, S. Leal, P. Parisi, C. Romero, W. Smith, A. Thaman, S. Warren, N. Yadav, Unity Perception: Generate Synthetic Data for Computer Vision (2021)

I. Brilakis et al., Toward automated generation of parametric BIMs based on hybrid video and laser scanning data. Adv. Eng. Inf. Elsevier **24**(4), 456–465 (2010). https://doi.org/10.1016/J.AEI. 2010.06.006

I. Brilakis et al., *Built environment digital twinning*, December 2019 (2019), pp. 17–18. https://asp ace.repository.cam.ac.uk/handle/1810/318329. Accessed 4 Aug 2022

P. Bukaty, *The California Consumer Privacy Act (CCPA): An Implementation Guide/Preston Bukaty* (IT Governance Publishing, Ely, Cambridgeshire, United Kingdom, 2019)

H. Caesar et al., Nuscenes: a multimodal dataset for autonomous driving, in *Proceedings of the IEEE Computer Society Conference on Computer Vision and Pattern Recognition* (IEEE Computer Society, 2020), pp. 11618–11628. https://doi.org/10.1109/CVPR42600.2020.01164

CEN: EN ISO 41011:2017. Facility management – vocabulary (2017)

I.G. Dino et al., Image-based construction of building energy models using computer vision. Autom. Constr. Elsevier B.V. **116** (2020). https://doi.org/10.1016/j.autcon.2020.103231

M.P. Gallaher et al., Cost analysis of inadequate interoperability in the U.S. capital facilities industry. Nist (2004), pp. 1–210. papers2://publication/uuid/69C8B354-4830-4874-929E-ACBCC00E3204

A. Geiger, P. Lenz, R. Urtasun, Are we ready for autonomous driving? the KITTI vision benchmark suite, in *Proceedings of the IEEE Computer Society Conference on Computer Vision and Pattern Recognition* (2012), pp. 3354–3361. https://doi.org/10.1109/CVPR.2012.6248074

Y. Hong et al., Synthetic data generation using building information models. Autom. Constr. Elsevier **130**, 103871 (2021). https://doi.org/10.1016/j.autcon.2021.103871

X. Hou, Y. Zeng, J. Xue, Detecting structural components of building engineering based on deep-learning method. J. Constr. Eng. Manag. American Society of Civil Engineers (ASCE) **146**(2) (2020). https://doi.org/10.1061/(ASCE)CO.1943-7862.0001751

G. Huang et al., Densely connected convolutional networks, in *Proceedings - 30th IEEE Conference on Computer Vision and Pattern Recognition, CVPR 2017*, January 2017 (Institute of Electrical and Electronics Engineers Inc., 2017), pp. 2261–2269. https://doi.org/10.1109/CVPR.2017.243

ISO, ISO 15686-3:2002 buildings and constructed assets—service life planning—part 3: performance audits and reviews (2002)

ISO: ISO 15686-1:2011 buildings and constructed assets—service life planning—part 1: general principles and framework (2010)

ISO, 'ISO 15686-7:2017 buildings and constructed assets—service life planning—part 7: performance evaluation for feedback of service life data from practice (2017)

W. Li et al., Interiornet: mega-scale multi-sensor photo-realistic indoor scenes dataset, in *British Machine Vision Conference 2018, BMVC 2018* (2019). https://interiornetdataset.github.io. Accessed 9 Aug 2022

Y. Li et al., PointCNN: convolution on X-transformed points. Adv. Neural Inf. Process. Syst. **31** (2018)

T.Y. Lin et al., Microsoft COCO: common objects in context, in *Lecture Notes in Computer Science (including subseries Lecture Notes in Artificial Intelligence and Lecture Notes in Bioinformatics)* (Springer, 2014), pp. 740–755. https://doi.org/10.1007/978-3-319-10602-1_48

M. Mahmoud et al., Automated BIM generation for large-scale indoor complex environments based on deep learning. Autom. Constr. Elsevier **162**, 105376 (2024). https://doi.org/10.1016/j.autcon.2024.105376

N. Mayer et al., What makes good synthetic training data for learning disparity and optical flow estimation? Int. J. Comput. vis. Springer New York LLC **126**(9), 942–960 (2018). https://doi.org/10.1007/s11263-018-1082-6

L. Nan, K. Xie, A. Sharf, A search-classify approach for cluttered indoor scene understanding. ACM Trans. Graph. ACMPUB27 New York, NY, USA (2012). https://doi.org/10.1145/2366145.2366156

M. Neuhausen, P. Herbers, M. König, Using synthetic data to improve and evaluate the tracking performance of construction workers on site. Appl. Sci. (Switzerland). MDPI AG **10**(14) (2020). https://doi.org/10.3390/app10144948

F.E. Nowruzi et al., How much real data do we actually need: analyzing object detection performance using synthetic and real data (2019). https://doi.org/10.48550/arxiv.1907.07061

Y. Pan et al., Void-growing: a novel Scan-to-BIM method for manhattan world buildings from point cloud, in *Proceedings of the 2021 European Conference on Computing in Construction*. ETH, vol. 2 (2021), pp. 312–321. https://doi.org/10.35490/EC3.2021.162

Y. Pan et al., Enriching geometric digital twins of buildings with small objects by fusing laser scanning and AI-based image recognition. Autom. Constr. Elsevier **140**, 104375 (2022). https://doi.org/10.1016/J.AUTCON.2022.104375

S. Paneru, I. Jeelani, Computer vision applications in construction: current state, opportunities & challenges. Autom. Constr. Elsevier **132**, 103940 (2021). https://doi.org/10.1016/J.AUTCON.2021.103940

X. Peng et al., Learning deep object detectors from 3D models, in *Proceedings of the IEEE International Conference on Computer Vision* (2015), pp. 1278–1286. https://doi.org/10.1109/ICCV.2015.151

Y. Peng et al., A hybrid data mining approach on BIM-based building operation and maintenance. Build. Environ. Pergamon **126**, 483–495 (2017). https://doi.org/10.1016/j.buildenv.2017.09.030

C.R. Qi et al., PointNet: deep learning on point sets for 3D classification and segmentation, in *Proceedings - 30th IEEE Conference on Computer Vision and Pattern Recognition, CVPR 2017* (Institute of Electrical and Electronics Engineers Inc., 2017a), pp. 77–85. https://doi.org/10.1109/CVPR.2017.16

C.R. Qi et al., PointNet++: deep hierarchical feature learning on point sets in a metric space. Adv. Neural Inf. Process. Syst. **30** (2017b)

L. Rampini, F.R. Cecconi, Artificial intelligence in construction asset management: a review of present status, challenges and future opportunities. ITcon **27**(43), 884–913 (2022). https://doi.org/10.36680/J.ITCON.2022.043, http://www.itcon.org/2022/43

J. Redmon et al., You only look once: unified, real-time object detection, in *Proceedings of the IEEE Computer Society Conference on Computer Vision and Pattern Recognition* (2016). https://doi.org/10.1109/CVPR.2016.91

M. Roberts et al., Hypersim: a photorealistic synthetic dataset for holistic indoor scene understanding (Institute of Electrical and Electronics Engineers (IEEE), 2020), pp. 10892–10902. https://doi.org/10.48550/arxiv.2011.02523

O. Russakovsky et al., ImageNet large scale visual recognition challenge. Int. J. Comput. vis. Springer, New York LLC **115**(3), 211–252 (2015). https://doi.org/10.1007/s11263-015-0816-y

H. Salehi, R. Burgueño, Emerging artificial intelligence methods in structural engineering. Eng. Struct. **171**, 170–189 (2018). https://doi.org/10.1016/j.engstruct.2018.05.084

S. Song, S.P. Lichtenberg, J. Xiao, SUN RGB-D: A RGB-D scene understanding benchmark suite, in *Proceedings of the IEEE Computer Society Conference on Computer Vision and Pattern Recognition* (IEEE Computer Society, 2015), pp. 567–576. https://doi.org/10.1109/CVPR.2015.7298655

T. Sutjaritvorakul, A. Vierling, K. Berns, Data-driven worker detection from load-view crane camera, in *Proceedings of the 37th International Symposium on Automation and Robotics in Construction, ISARC 2020: From Demonstration to Practical Use - To New Stage of Construction Robot.* (International Association on Automation and Robotics in Construction (IAARC), 2020), pp. 864–871. https://doi.org/10.22260/isarc2020/0119

J. Tremblay et al., Training deep networks with synthetic data: bridging the reality gap by domain randomization, in *IEEE Computer Society Conference on Computer Vision and Pattern Recognition Workshops* (IEEE Computer Society, June 2018, 2018), pp. 1082–1090. https://doi.org/10.48550/arxiv.1804.06516

P. Voigt, A. von dem Bussche, *The EU General Data Protection Regulation (GDPR)* (Springer International Publishing, 2020). https://doi.org/10.1093/oso/9780198826491.001.0001

C. Wang, Y.K. Cho, C. Kim, Automatic BIM component extraction from point clouds of existing buildings for sustainability applications. Autom. Constr. Elsevier **56**, 1–13 (2015). https://doi.org/10.1016/j.autcon.2015.04.001

R. Wang, L. Xie, D. Chen, Modeling indoor spaces using decomposition and reconstruction of structural elements. Photogramm. Eng. Remote. Sens. Am. Soc. Photogramm. Remote. Sens. **83**(12), 827–841 (2017). https://doi.org/10.14358/PERS.83.12.827

Y. Wei, B. Akinci, A vision and learning-based indoor localization and semantic mapping framework for facility operations and management. Autom. Constr. Elsevier **107**, 102915 (2019). https://doi.org/10.1016/J.AUTCON.2019.102915

Y. Wei, B. Akinci, Synthetic image data generation for semantic understanding in everchanging scenes using BIM and Unreal engine. In *Computing in Civil Engineering 2021* **115**, (pp. 934–941) (n.d.). https://doi.org/10.1061/9780784483893

Z.Q. Zhao et al., Object detection with deep learning: a review. IEEE Trans. Neural Netw. Learn. Syst. Institute of Electrical and Electronics Engineers Inc. **30**(11), 3212–3232 (2018). https://doi.org/10.48550/arxiv.1807.05511

# Chapter 6
# Industry 5.0 in Construction: Towards a More Human-Centric and Ethical AI

The construction industry has increasingly embraced the advancements of digital technologies under the umbrella of Industry 4.0, giving rise to the concept of "Construction 4.0." This goes beyond simply upgrading traditional methods; it represents a paradigm shift as the sector adopts and adapts Industry 4.0 technologies such as Building Information Modeling (BIM), 3D printing, the Internet of Things (IoT), robotics, and big data to drive digitalization, automation, and the expanded use of information and communication technologies (ICT) (Regona et al. 2022). In practice, Construction 4.0 enables the capture, storage, processing, and communication of data, fostering collaboration and efficiency. This results in optimized supply chains, real-time process control through virtual models, and proactive problem-solving, transforming construction into a technology-driven industry (Musarat et al. 2021). The integration of cyber-physical systems, such as combining Building Information Modelling (BIM) with IoT for real-time monitoring and decentralized decision-making, or Artificial Intelligence (AI) ability to process huge enormous data generated in construction projects, exemplifies the potential of Industry 4.0 technologies to revolutionize the construction sector (Irani and Kamal 2014).

This transition is crucial for the construction industry to evolve into a more efficient and innovative sector, aligning with global trends in digital transformation. However, despite its promise, Construction 4.0 has faced slower adoption, particularly in developing countries where traditional, labor-intensive practices remain dominant. As artificial intelligence (AI) continues to penetrate this historically manual sector, it introduces significant ethical challenges and potential biases that must be addressed. The main objectives of Industry 4.0 technologies are to enhance productivity and efficiency through automation and the integration of cyber-physical systems (Maskuriy et al. 2019). Yet, in the drive to achieve these goals, the ethical, moral, and social impacts of technology applications are sometimes overlooked (Khodabakhshian 2024). The construction industry, which uniquely depends on both automated systems

By Ania Khodabakhshian.

© The Author(s), under exclusive license to Springer Nature Switzerland AG 2024
F. Re Cecconi et al., *Building Tomorrow: Unleashing the Potential of Artificial Intelligence in Construction*, PoliMI SpringerBriefs,
https://doi.org/10.1007/978-3-031-77197-2_6

and human labor, presents a complex environment where the ethical implications of AI are especially critical. Concerns such as job displacement, data privacy, transparency in AI decision-making processes, and algorithmic biases are central to the discussion on AI in construction (Weber-Lewerenz 2021).

In contrast, Industry 5.0 shifts the focus towards human-centricity, emphasizing the collaboration between humans and intelligent machines to achieve more personalized, sustainable, and resilient production processes (Zizic et al. 2022). This shift seeks to create a balanced coexistence between humans and machines in industrial environments. Industry 5.0 reintroduces the human element into production systems, promoting ethical considerations, creativity, and social well-being alongside technological advancement (Madsen and Slåtten 2023). In this new era, technologies such as collaborative robots (cobots), AI, and IoT work together to create more resilient, sustainable, and user-friendly construction environments (Maddikunta et al. 2021; Marinelli 2023).

Despite its benefits, the Industry 5.0 concept faces challenges, such as the demand for advanced technical skills and the integration of human-centric technologies into existing systems. Current research indicates that while Industry 5.0 is still in its early stages, its potential applications in construction—such as real-time environmental monitoring and waste management—show promise for transforming the sector. These advancements could make construction more sustainable and resilient, aligning with Industry 5.0's broader goal of creating a human-centered technological ecosystem (Nahavandi 2019; Tunji-Olayeni et al. 2024a).

The Industry 5.0 framework offers a promising perspective for addressing the ethical concerns associated with AI and other digital technologies in construction. It places human values and well-being at the forefront of technological integration, promoting the concept of "responsible innovation." This approach ensures that the development and deployment of digital technologies are guided by ethical standards and societal values (Nahavandi 2019; Tunji-Olayeni et al. 2024b). Unlike Industry 4.0's focus on automation and efficiency, Industry 5.0 aims to create AI systems that adhere to ethical norms such as transparency, accountability, and fairness. This is particularly important in construction, where AI is increasingly being used for decision-making, risk management, and project optimization. By incorporating ethical AI frameworks, Industry 5.0 ensures that AI applications are inclusive, non-discriminatory, and aligned with human-centered values, mitigating risks related to bias, job displacement, and privacy concerns.

For this transition to succeed, AI systems must demonstrate that they serve humanity, transforming workers' roles rather than replacing them, promoting human values and ethical norms, and building trust among stakeholders (Emaminejad and Akhavian 2022). Establishing trust, especially in a labor-intensive industry like construction, requires extensive research on ethical values and potential obstacles (Emaminejad et al. 2015). These challenges include biases and discrimination caused or exacerbated by AI, fears of job loss and privacy violations, and conflicts of autonomy and interests between humans and machines in decision-making processes. These topics will be explored in detail in this chapter.

After identifying the potential harms and biases, the concept of "Responsible AI" must be applied—a methodology that translates ethical principles such as fairness, transparency, explainability, accountability, safety, reliability, security, and privacy into practical, measurable metrics for industrial AI applications. The Responsible AI process involves two key phases: planning and development, and deployment and monitoring, which will be discussed in the following sections (Askell et al. 2019; Fjeld et al. 2020).

The planning and development phase includes implementing bias mitigation strategies and frameworks in AI system design. The deployment and monitoring phase involves establishing a governance system to actively supervise the effectiveness of these strategies. In an industry as fragmented as construction, with numerous stakeholders, clear guidelines and metrics must be established to manage the ethical implications of AI. Ultimately, ethical AI metrics can be defined and included as key performance indicators (KPIs) for the entire AI application process. This will motivate companies to contribute actively to minimizing ethical issues and biases in AI. This chapter aims to draw attention to the most important factors during the development and deployment of ethical AI applications.

## 6.1  Ethics of AI

Ethics, a set of moral principles, guide individuals and organizations in fostering trust and promoting positive societal outcomes (*Marriam-Webster Dictionary*, no date). As AI and robotics increasingly permeate various aspects of life, managing their societal impacts, decision-making capabilities, and levels of authority becomes crucial. These technologies raise complex ethical questions that go beyond legal requirements, emphasizing the need for morally sound behavior to build societal trust (Kuipers 2020).

AI ethics encompasses the broader concerns related to the development, deployment, and societal impact of AI systems, focusing on the moral responsibilities of those who design, build, and use these technologies. It addresses issues such as data bias, privacy, unemployment, and, as AI advances, even the potential consideration of robot rights. In contrast, ethical AI focuses on the behavior and decision-making processes of AI systems themselves, ensuring that they act responsibly and adhere to ethical principles (Siau and Wang 2020).

Artificial Intelligence (AI) mimics human cognitive functions, such as perception, reasoning, learning, and decision-making, by leveraging machine learning (ML) algorithms to analyze data and develop decision-making models. However, these models can perpetuate and even amplify existing biases in the underlying data, posing significant challenges, particularly in industries like construction (Rai and Sarker 2019; Akter et al. 2022).

To mitigate such biases in AI applications, ethics must play a key role at multiple levels, from engineers designing these systems to policymakers and society at large. Ensuring that AI systems not only perform effectively but also align with societal values and expectations is essential. Ethical considerations influence every stage of AI system design, from defining the system's purpose to implementing the finer details of its function. This ethical foundation is critical in shaping the kind of society we aim to create (Bartneck et al., no date). However, developing comprehensive ethical standards that AI systems can understand and follow is complex, as emotional and moral concepts like harm, bias, and discrimination are not easily translated into machine language. Bridging the gap between AI programmers and ethical standards creators is essential to ensure that AI systems behave ethically. Researchers and practitioners must deepen their understanding of ethical principles to apply them effectively in AI development (Wang and Siau 2019).

Achieving ethical AI requires more than just principles; it demands robust governance frameworks throughout the AI lifecycle—from data selection to deployment and ongoing management. This involves establishing ethics boards and implementing rigorous quality assurance processes to guide the ethical development and use of AI technologies (Eitel-Porter 2021). AI and robotics also pose several ethical risks that must be managed in a globalized economy, including reputational risks from biased AI systems, legal risks related to anti-competitive practices, environmental risks from AI failures, and social risks such as increased inequality and privacy concerns. To address these risks, companies should integrate ethical values into their strategies and exceed safety regulations when necessary (Bartneck et al., no date).

## 6.2   Harms and Biases of AI

### 6.2.1   Data Bias

AI's capacity to process vast amounts of data and make autonomous decisions is both its greatest strength and a potential vulnerability, particularly in a complex, human-centric industry like construction (Weber-Lewerenz 2021). In this context, the quality of data gathering becomes even more critical, as the construction industry often deals with poor and unstructured data (Arroyo et al. 2021). This is a key concern, given that around 85% of AI projects may produce inaccurate results due to biases in the data, algorithms, or the teams managing them (Gartner 2018).

A major issue is algorithmic bias, which can arise from unrepresentative data used to train models (Syam and Sharma 2018), whether due to socio-cultural contexts, poor data collection, or the improper selection and design of algorithms (Balducci and Marinova 2018). AI has the potential to replicate and even amplify existing disparities and biases found in the data, such as gender or racial biases. For instance, in a male-dominated industry like construction, AI systems trained on biased data could reinforce gender and racial inequalities in hiring practices.

Moreover, the use of unstructured data—such as text documents, images, videos, emails, and voice messages—comprising roughly 80% of all business data, adds another layer of complexity. When data is unlabeled and unstructured, unsupervised machine learning (ML) algorithms are typically used to detect patterns and under-lying structures for clustering. However, this approach can exacerbate biases, as unstructured data can be interpreted and labeled in multiple ways (Akter et al. 2022). Human biases introduced during manual labeling can result in skewed samples and distorted outcomes. Additionally, qualitative and simplistic labeling—such as cate-gorizing risks as "high" or "low"—can make the learning process less interpretable for machines. Biases can also enter through the selection of sensitive features or by masking protected characteristics with new ones, leading to lower accuracy for certain groups (Mujtaba and Mahapatra 2019).

While it's tempting to collect as much data as possible to avoid missing anything, it's essential to ask whether AI can adequately account for all political, moral, and social factors, as well as inherent biases. Additionally, phenomena like groupthink—where members avoid disagreement to maintain consensus—can suppress diverse viewpoints. There is concern that AI might exacerbate this issue by making it easier to bypass difficult discussions, potentially sidelining critical ethical considerations (Arroyo et al. 2021).

## 6.2.2 Model Bias

Bias in AI algorithms can arise from multiple sources, leading to flawed outcomes that fail to accurately capture the true causal relationships within the data. The perfor-mance of any AI model is highly dependent on the quality of the input data and the model's design and specifications (Walsh et al. 2021). When the dataset is unrepre-sentative, the model lacks relevant attributes, is poorly specified, or is inappropriately applied to a specific context, the resulting outcomes are likely to be unreliable and potentially discriminatory (Khodabakhshian 2024).

In machine learning models, for instance, bias occurs when the mathematical models are inadequately specified, causing them to miss critical interrelationships and causalities between input features and output variables. This can have adverse effects on certain groups, such as female workers, especially if the model's design fails to account for variance or overgeneralizes its findings (Rozado 2020).

Bias can also result from the improper application of algorithms to specific contexts. For example, in risk management scenarios where probabilistic models are needed to handle inherent uncertainties, using deterministic models instead can produce unrealistic outputs, rendering the models ineffective for managing risks (Khodabakhshian et al. 2023). However, it is important to distinguish this issue from contextual bias, which arises from (a) cultural bias—when the model fails to capture values, norms, beliefs, and behaviors that individuals learn from society due to their difficult-to-quantify nature, (b) social bias—when the model dispropor-tionately impacts individuals from smaller social groups, making their social roles and status more visible, and (c) Personal bias—referring to how algorithms treat individuals differently based on personal characteristics such as gender, age, and personality (Akter et al. 2022). However, While AI promises enhanced personalization, it also risks reducing human interaction and promoting social isolation. Over-personalization can limit exposure to diverse viewpoints, potentially contributing to social polarization (Leslie 2019).

Methodological biases frequently emerge at various stages of the machine learning lifecycle—from problem conceptualization to ongoing maintenance—often due to developers' lack of experience or knowledge about the specific application context. These biases can lead to discriminatory outcomes, correlation fallacies, and over-generalization (Lorenzoni et al. 2021; Walsh et al. 2021). Addressing these issues requires thorough error analysis, adjusting model features, and selecting appropriate model architectures, though these solutions may introduce challenges like increased variance and higher computational costs.

Ensuring that AI systems are developed and deployed in ways that are fair, transparent, and accountable is essential. This includes utilizing bias detection tools, involving diverse teams in AI development, and implementing rigorous testing protocols. By taking these steps, the risk of embedding and amplifying existing biases in algorithmic decision-making can be mitigated, leading to more equitable outcomes in AI applications such as recruitment, safety assessments, and project management (Mujtaba and Mahapatra 2019). Such mitigation strategies are vital to gaining public trust and acceptance of the digital technologies, as poor data management, negligent design, and irresponsible deployment can result in AI systems that are unreliable, unsafe, or produce subpar outcomes, and leading to inefficient use of resources (Leslie 2019).

### 6.2.3  Bias Mitigation Methods

Bias in AI models can be mitigated through three primary methods. The first is pre-processing, which involves editing the dataset before training by modifying features and labels in accordance with fairness criteria. This can include removing protected attributes or adjusting biased features to ensure a more balanced dataset, ultimately reducing bias in the input data itself. The second approach is in-processing or optimization, which focuses on adjusting the model during training to meet fairness standards by applying constraints to the classification objectives. While this method can enhance fairness, it may affect the model's accuracy and can be challenging to implement if the classifier is developed externally or outsourced. The third method is post-processing or counterfactual analysis, where the model's outcomes are adjusted after training to align with fairness objectives. This is done by setting thresholds or introducing counterfactual explanations, allowing fairness criteria to be met without altering the classifier itself. This approach can improve the system's perceived fairness while maintaining the model's integrity (Mujtaba and Mahapatra 2019).

### 6.2.4 Job Disruption and Dynamic Change

AI and digital technologies are poised to significantly enhance productivity in the construction industry and contribute to global economic growth. However, they also present challenges, particularly in terms of job displacement, as repetitive tasks become increasingly automated and human roles are replaced by machines (Torresen 2018). For instance, advancements in large language models like ChatGPT raise concerns about job security in areas such as contract management within the construction sector (Prieto et al. 2023). While up to 375 million workers—about 14 percent of the global workforce—may need to switch occupations by 2030, this shift demands higher education, new skills, and adaptability to the digital nature of emerging jobs. These requirements often create resistance to change among industry professionals, reducing the acceptance rate of digital technologies in traditional sectors like construction (Mckinsey Global Institute 2017).

As AI and automation technologies advance, they can perform tasks traditionally carried out by both skilled and unskilled laborers, such as bricklaying, welding, and even project management. This presents a significant risk of economic and social disruption within the construction industry. While these technologies offer potential cost savings and efficiency improvements, they also threaten employment, particularly in sectors dependent on manual labor. This raises ethical concerns about the responsibility of companies to support affected workers through retraining and transition programs (Weber-Lewerenz 2021).

Despite the potential for disruption, history has shown that technology typically creates more jobs than it eliminates, driving demand for new roles, improving productivity, and fostering a better work-life balance (Economist 2016). McKinsey predicts that only 0–33% of jobs may be displaced, with the potential for new jobs to be created in greater numbers (McKinsey Global Institute 2016). In construction, roles that are more repetitive and require less specialized knowledge—such as certain site activities or detailed design tasks—are more susceptible to automation compared to strategic roles like project management. However, companies must ensure that these technologies augment human workers rather than replace them entirely. For example, collaborative robots (cobots) are designed to improve worker safety and well-being without eliminating human labor. This approach balances technological progress with social responsibility, ensuring that the benefits of AI and digital technologies are distributed equitably without causing negative societal impacts (Lorenzini et al. 2023).

Policymakers and business leaders must invest in worker transitions and promote economic dynamism to ensure that job growth outpaces job losses. In the construction and infrastructure sectors, where capital investments of around $3.3 trillion per year are needed to address infrastructure gaps and housing shortages, labor demand could be significantly boosted—potentially creating up to 80 million jobs by 2030 (McKinsey Global Institute 2016).

AI and digital technologies also have the potential to reshape the nature and dynamics of jobs, as well as the work-life balance of workers over the long term.

As machines take over repetitive tasks, human workers will have more time to focus on complex and strategic activities, such as deciding which projects to prioritize and invest in (Nam 2014). However, increased automation could blur the boundaries between work and personal life, making it more challenging to distinguish work hours from leisure time. While automation may reduce the physical demands of jobs, it could also lead to longer working hours, as technology enables work to be done from anywhere at any time (Santhosh et al. 2023).

Another concern is the potential loss of specialized skills. As AI provides quick access to information and solutions, human workers may become less motivated to learn and develop specialized knowledge, increasingly relying on machines. For instance, the rise of tools like ChatGPT poses risks to the quality of learning for students and interns, who may turn to AI for answers rather than acquiring knowledge through personal effort. The increasing reliance on robots and information technology could also reduce the need for human cognitive skills, such as memory, reasoning, and decision-making. This may make it difficult to distinguish between expert knowledge and unverified information available online or through AI models (Orrù et al. 2023).

AI's influence on the construction industry also introduces motivational and ethical challenges. According to Self-Determination Theory (SDT), motivation is driven by the fulfillment of autonomy, competence, and relatedness (Deci et al. 2024). AI could significantly impact these factors by reducing autonomous decision-making and hindering opportunities for learning and personal development. As AI takes on more decision-making roles, collaboration within project teams may be affected, potentially diminishing motivation and overall performance. Addressing these ethical challenges will be crucial to ensuring that the integration of AI into the construction industry benefits both the workforce and society at large (Schöttle 2020).

## 6.3   Ethical AI Metrics for Model Design and Development

Researchers have introduced concepts such as moral autonomous systems, ethical controllers, and Responsible AI to differentiate between the typically amoral decision-making mechanisms of AI agents and their moral frameworks, which establish the ethical context for their actions. Decision-making policies in AI should be evaluated and prioritized according to this ethical context, with the highest-ranked option selected for action. Autonomous systems that fail to meet ethical standards are deemed impractical or unrealizable (Svegliato et al. 2020; Mezgár and Váncza 2022).

Defining ethical AI metrics in the construction industry requires a multifaceted approach that addresses the unique challenges and demands of the sector. These metrics must be integrated into the design and development of AI systems to ensure fairness, transparency, and respect for human rights. A comprehensive evaluation framework is essential, one that considers both technical and social dimensions (European Parliament 2019). This includes creating performance indicators that measure not only the efficiency and productivity improvements delivered by AI but also its social impact, such as effects on employment and worker safety (Pillai and Matus 2020).

As Torresen (2018) suggests, addressing the ethical challenges of AI and robotics requires engineers to prevent misuse and ensure human oversight. Autonomous AI systems must be capable of ethical decision-making to minimize the risk of undesirable behavior. To this end, AI systems need a set of ethical metrics as a benchmark for evaluating the moral implications of their actions (Torresen 2018).

The European Union's AI Act encourages the development of AI that is transparent, accountable, and aligned with EU values, including respect for human autonomy, harm prevention, fairness, and explainability. The Act mandates that AI systems, particularly high-risk ones, be designed to be transparent and explainable. Users must be informed when interacting with AI and understand how decisions that affect them are made (European Parliament 2019). Furthermore, Weber-Lawerenz (2021) proposed the concept of Corporate Digital Responsibility (CDR), which ensures that AI systems are developed with an ethics-by-design approach, adhering to ethical principles throughout the AI lifecycle. By integrating these ethical considerations, companies can build trust and ensure that AI-driven innovations positively contribute to industry progress (Weber-Lewerenz 2021). The European Commission has published ethics guidelines for trustworthy AI, defining four ethical principles and seven requirements, which are used to develop an assessment list for trustworthy AI (ALTAI) (European Commission 2020).

Monitoring the use of AI in construction is essential to ensure it does not exacerbate existing inequalities or introduce new risks, particularly concerning worker treatment and the equitable distribution of benefits. By adopting a holistic, interdisciplinary approach to defining ethical metrics, the construction industry can better navigate the challenges posed by AI and ensure that technological advancements lead to sustainable and inclusive growth (Pillai and Matus 2020).

A comprehensive list of ethical AI metrics identified by previous researchers is presented in Table 6.1.

## 6.4 Developing Ethical AI Systems

As AI systems become increasingly complex, embedding ethical decision-making capabilities within them is essential (Dennis et al. 2015). This requires collaboration between programmers and ethicists to create autonomous machines that adhere to ethical guidelines. Wallach and Allen (2009) explored the development of artificial moral agents, proposing three approaches: formal ethical reasoning, machine learning based on examples, and simulations to observe the outcomes of different ethical strategies (Wallach and Allen 2009).

When software replaces human judgment, it must follow key principles such as accountability, transparency, robustness, and predictability. Developers should

**Table 6.1** Ethical AI metrics and components

| Metrics | Description | Application in construction industry | References |
|---|---|---|---|
| Data Privacy and Security (level of surveillance) | AI systems often use personal data without proper consent, raising significant privacy concerns and potentially infringing on individual rights, specially in high-risk systems. Protecting sensitive information is essential for maintaining trust in AI systems, especially as technological advancements blur the line between privacy and data usage | In the construction industry, where AI relies heavily on large datasets, including sensitive personal and organizational data, robust cybersecurity measures are crucial to address these concerns and insure an ethical level of surveillance of workers | (Emaminejad et al. 2015; Future of Life Institute 2017; UNESCO 2017; Floridi et al. 2018; IAPP (International Association of Privacy Professionals) 2018; The Public Voice 2018; European Commission 2019b; European Parliament 2019; IEEE (Institute of Electrical and Electronics Engineers) 2019; Leslie 2019; Eitel-Porter 2021; Emaminejad and Akhavian 2022; Liu et al. 2022; Liang et al. 2023) |
| Data and model Transparency and explainability | Explainability and transparency are essential for building trust in AI systems. Users must understand AI model's workflow, capabilities, and limitations to ensure accountability and prevent bias. Explainable AI (XAI) transforms these models into more interpretable systems, supporting clear decision-making and human oversight, and addressing the limitations of "black box" models | This finds special importance in industries like construction where AI makes critical decisions that can have far-reaching consequences | (Emaminejad et al. 2015; Future of Life Institute 2017; Floridi et al. 2018; IAPP (International Association of Privacy Professionals) 2018; The Public Voice 2018; Weinberger 2018; European Commission 2019b, 2020; European Parliament 2019; IEEE (Institute of Electrical and Electronics Engineers) 2019; Leslie 2019; Eitel-Porter 2021; Thiebes et al. 2021; Emaminejad and Akhavian 2022; Liu et al. 2022; Liang et al. 2023) |

(continued)

**Table 6.1**   (continued)

| Metrics | Description | Application in construction industry | References |
|---|---|---|---|
| Respect to human autonomy | AI systems must respect human autonomy by supporting, not overriding, human decision-making, especially in complex scenarios involving general intelligence AI, where the system is entrusted with making major decisions. Proper oversight mechanisms, such as human-in-the-loop approaches, are crucial to preserving human agency, accountability, and rights | This maintains ethical integrity and accountability, and limits the potential conflicts of interest between human and machine agents' decision | (Future of Life Institute 2017; UNESCO 2017; Floridi et al. 2018; European Commission 2019b; Leslie 2019; Thiebes et al. 2021; Liang et al. 2023) |
| Acceptance and Trust | Building trust between human workers and AI systems is crucial for effective collaboration. Effective Human-technology interaction requires acceptance from public, the prerequisite of which is trust in the system and its decisions. Trust calibration is essential to balance trust, preventing both overtrust and undertrust in AI and robotics | Gaining trust and acceptance from construction professionals will greatly increase AI application rate in the change-resistant industry | (European Commission 2019b, 2020; Emaminejad and Akhavian 2022; Liang et al. 2023) |

(continued)

**Table 6.1** (continued)

| Metrics | Description | Application in construction industry | References |
|---|---|---|---|
| Reliability and Safety (Technical performance and robustness) | Reliability involves ensuring consistent, failure-free operation, while safety addresses both technical (avoiding accidents) and psychological (viewing AI as a collaborator, not a threat) aspects. AI systems need to be safe, ensuring a fall back plan in case something goes wrong to minimize the unintentional harms | In construction industry, the undetected errors, failures and misuses resulting from wrong and unexplained decisions of AI poses great risks to the projects and workforce | (Emaminejad et al. 2015; Future of Life Institute 2017; Floridi et al. 2018; European Commission 2019b, 2020; Liang et al. 2023) |
| Liability And accountability | A clear chain of command is crucial for addressing biased or erroneous AI outcomes, ensuring accountability and prompt corrective action. Implementing auditability is key to assessing algorithms and design processes, especially in critical applications. Furthermore, Addressing the legal implications and responsibility for AI and robotic actions in the workplace is vital | Errors in algorithmic decision-making in construction create a "liability puzzle," making it unclear who—developers, manufacturers, or contractors—is ethically and legally responsible for harmful outcomes | (Future of Life Institute 2017; UNESCO 2017; Floridi et al. 2018; IAPP (International Association of Privacy Professionals) 2018; The Public Voice 2018; European Commission 2019b, 2020; IEEE (Institute of Electrical and Electronics Engineers) 2019; Pillai and Matus 2020; Eitel-Porter 2021; Liang et al. 2023) |
| Non-maleficence | AI systems must be designed to avoid causing harm to people, animals, or property. Protecting against such harm is crucial for ensuring individual safety and maintaining the company's reputation | Harms in construction can range from physical harms and accidents to discrimination hiring processes automated by AI | (Floridi et al. 2018; European Commission 2019b, 2020; Thiebes et al. 2021) |

(continued)

**Table 6.1** (continued)

| Metrics | Description | Application in construction industry | References |
|---|---|---|---|
| Beneficence | AI should be designed with the goal of doing good and improving lives. Rooted in bioethics, this principle ensures that the benefits of AI outweigh any potential harms, making these systems trustworthy and fair | Automating repetitive tasks by AI, or assisting workers in physically intensive works in dangerous construction sites by robots, such technologies contribute to mental and physical health of the workforce | (Floridi et al. 2018; Thiebes et al. 2021) |
| Justice and Fairness | Defining justice in AI is complex, but clear regulations and ethical norms can guide AI to make fair decisions. Eliminating bias and ensuring consistent, equitable outcomes across different subgroups of society are crucial for maintaining trust | AI systems must avoid discrimination, promote diversity, and be accessible to all, involving stakeholders throughout their lifecycle in construction companies | (UNESCO 2017; Floridi et al. 2018; IAPP (International Association of Privacy Professionals) 2018; The Public Voice 2018; European Commission 2019b, 2020; Eitel-Porter 2021; Thiebes et al. 2021; Liu et al. 2022) |
| Social and environmental well-being | AI systems should benefit all human beings, including future generations. It must hence be ensured that they are sustainable and environmentally friendly. Moreover, they should take into account the environment, including other living beings, and their social and societal impact should be carefully considered | As sustainability is a current hype in construction, it is vital to make sure that AI is not contributing to greenwashing and does not cause long term social harms for short term economical gains | (Future of Life Institute 2017; UNESCO 2017; IAPP (International Association of Privacy Professionals) 2018; The Public Voice 2018; European Commission 2019b, 2020; IEEE (Institute of Electrical and Electronics Engineers) 2019; Liu et al. 2022) |

ensure that these systems can be inspected and corrected when errors occur, reducing the risk of unethical consequences.

Research in robot ethics has led to the creation of self-aware robots capable of making ethical decisions in real-world scenarios. For instance, robots like Nao have been programmed to balance patient safety, autonomy, and ethical behavior in healthcare settings, showcasing progress toward ethical, autonomous systems (Winfield et al. 2014).

To ensure a high level of protection for public interests—particularly concerning health, safety, and fundamental rights—common regulations for high-risk AI systems must be established. These rules should be consistent with the Charter of Fundamental Rights of the European Union, non-discriminatory, and aligned with the Union's international trade commitments. Additionally, they should reflect the European Declaration on Digital Rights and Principles for the Digital Decade and the Ethics Guidelines for Trustworthy AI, developed by the High-Level Expert Group on Artificial Intelligence (AI HLEG) (European Commission 2019a, b).

## 6.5  Ethical AI Deployment and Governance

To successfully implement AI and robotics and gain the trust of both management and users, their value in improving workflows must be clearly demonstrated. Research in construction shows that perceived benefits in safety, efficiency, and cost savings play a significant role in technology adoption. Simplifying user interactions through intuitive interfaces and familiar technologies like BIM can further drive adoption, making AI and robotics more accessible and seamlessly integrated into traditional workflows. Workforce training and education are key facilitators of broader AI integration, enhancing human–machine interaction and improving oversight of AI models (Enaminejad and Akhavian 2022).

In construction, an ethical AI strategy should align with sustainability goals and promote ethical awareness throughout the AI lifecycle. Companies must balance ethical considerations with economic efficiency, adopt secure data platforms, and ensure digital transformation maximizes social, economic, and environmental benefits. Additionally, ethical AI deployment must address the industry's skills gap through improved training and collaboration. Organizations should take responsibility for the ethical use of these technologies to better track accountability, especially in systems with higher levels of intelligence (Weber-Lewerenz 2021).

AI can be categorized into weak (narrow) AI and strong AI based on its level of human supervision and intelligence. Weak AI, or artificial narrow intelligence (ANI), assists with specific tasks under human supervision, enhancing decision-making without replacing human authority and posing minimal risk to human motivation. Strong AI, including artificial general intelligence (AGI) and artificial superintelligence (ASI), matches or surpasses human cognitive abilities across various domains (Müller, no date). While AGI aims to reach human-level intelligence, ASI could exceed it, raising significant ethical concerns due to its potential to operate with minimal human intervention, disrupt job markets, and even claim a form of legal personality. Human oversight is crucial to maintaining ethical standards and ensuring harmful decisions by AI can be overridden (McAleenan 2020). According to Accenture, while 63% of leaders recognize the importance of monitoring AI systems, many are unsure how to do so, with about 60% requiring a human override of an AI system at least once a month (Kanioura 2019).

## *6.5.1 Governance Framework*

An ethical AI governance framework provides a systematic approach to ensure the development, deployment, and use of AI systems align with ethical principles, human rights, and societal values. This framework fosters trust, promotes transparency, and mitigates risks associated with AI. Organizations like the Future of Life Institute, IEEE, and the European Commission emphasize key principles such as accountability, transparency, privacy, and human rights to guide responsible AI development (Future of Life Institute 2017; IEEE (Institute of Electrical and Electronics Engineers) 2019).

The Alan Turing Institute proposed an ethical framework with these components:

(a) SUM Values: Respect, Connect, Care, and Protect—principles for responsible data design and use.
(b) FAST Track Principles: Fairness, Accountability, Sustainability, and Transparency—ensuring ethical AI system design and deployment.
(c) PBG Framework: A process-based governance structure to operationalize ethical principles across AI project workflows (Leslie 2019).

Windfield and Jirotka (2018) proposed a roadmap for AI governance that integrates ethics, standards, regulation, responsible research, innovation, and public engagement to build public trust (Winfield and Jirotka 2018). Similarly, Wu et al. (2020) emphasized the role of ethics and governance in China's AI sustainable development, advocating for ethical guidelines and governance technologies to contribute to society (Wu et al. 2020). Mäntymäki et al. (2022) introduced the Hourglass Model, which helps organizations implement ethical AI principles and align their AI processes with regulatory frameworks like the European AI Act. This model addresses governance at environmental, organizational, and AI system levels, ensuring comprehensive oversight throughout an AI system's lifecycle. The Hourglass Model's AI system layer includes eight key components: A. AI system B. Algorithms C. Data operations D. Risk and impacts E. Transparency, explainability, and contestability F. Accountability and ownership G. Development and operations H. Compliance. This framework underscores the systemic nature of AI governance and serves as a foundation for organizational decision-makers to ensure social acceptability, mitigate risks, and harness the full potential of AI (Mäntymäki et al., no date).

## 6.5.2  AI Ethical Standards

Governments and standards organizations are increasingly developing regulations and guidelines to ensure the ethical use of AI. Since the 1980s, technical ethics has expanded to address the challenges posed by emerging technologies. The IEEE launched a global initiative in 2016 focusing on the ethics of autonomous and intelligent systems, prioritizing human well-being. Similarly, the British Standards Institution introduced guidelines for conducting ethical risk assessments in AI systems (Liang et al. 2023).

International bodies such as ISO and IEC are working on AI ethics standards. ISO published the "Overview of Trustworthiness in Artificial Intelligence," (International Organization for Standardization 2021), while the EU's CEN and CENELEC are developing AI standards aligned with the proposed EU AI regulation. In the U.S., NIST has established trustworthiness standards covering accuracy, explainability, safety, and security (National Institute of Standards and Technology (NIST) 2021).

The European AI Act categorizes AI systems based on risk levels:

(a)  Unacceptable risk: Systems like social scoring are banned.
(b)  High-risk AI: These systems are heavily regulated.
(c)  Limited-risk AI: These systems have lighter transparency requirements.
(d)  Minimal-risk AI: Generally unregulated, including video games and some generative AI applications (European Parliament 2019).

The AI Act establishes the European Artificial Intelligence Board to ensure consistent application across the EU, alongside the Digital Services Act (DSA) and General Data Protection Regulation (GDPR), which govern data privacy and security in AI systems. The EU's guidelines for trustworthy AI emphasize lawfulness, ethics, and robustness, and outline seven requirements: human agency, technical robustness, privacy, transparency, accountability, social well-being, and fairness (European Commission 2019a).

The European Commission's guidelines for trustworthy AI outline three key principles—lawfulness, ethics, and robustness—and seven requirements, including human agency, technical robustness, privacy, transparency, accountability, social and environmental well-being, and diversity and fairness, to ensure AI systems are trustworthy. The AI HLEG's Assessment List for Trustworthy Artificial Intelligence (ALTAI) is a tool designed to help organizations ensure their AI systems adhere to these principles. While ALTAI is broadly applied across sectors, its specific use in industries like construction has not been extensively documented (European Commission 2020). The EU's General Data Protection Regulation (GDPR) has set a global precedent for personal data protection, influencing AI regulations worldwide. As AI continues to impact areas like labor laws, global cooperation will be needed to develop effective, ethical regulations that benefit all, including developing countries (Weinberger 2018).

Globally, there is increasing interest in AI and robotics, driven by economic competitiveness. Countries like Canada, Japan, and China have launched AI strategies aimed at fostering innovation while ensuring ethical development. Public concerns, such as job displacement and privacy loss, have prompted a growing emphasis on ethics in AI policies. Germany and France, for example, include societal discussions and ethics-by-design in their AI strategies to address these concerns (Ajay Agrawal 2019).

In the construction industry, AI will need to adhere to ethical guidelines as it becomes more common. ISO standards, such as ISO 12100:2010, offer a framework for risk assessment in construction machinery design, prioritizing safety and responsible design. McAleenan et al. (2019) identified gaps in the ethical development of construction safety technologies, noting that while these technologies aim to enhance safety, they often neglect workers' rights. This challenge goes beyond Asimov's first law of robotics and requires AI systems to consistently prioritize human well-being.

The construction industry, regulated by ISO standards, will need to incorporate these ethical guidelines as AI systems become more common. Standards like ISO 12100:2010 provide a framework for risk assessment and reduction in construction machinery design, prioritizing safety and informed design decisions. This approach ensures that AI development in the construction industry, and beyond, remains responsible, ethical, and aligned with societal well-being and safety (Pillai and Matus 2020; International Organization for Standardization 2021). McAleenan et al. (2019) highlighted a gap in the ethical development of OSH-monitoring technologies, noting that while these technologies aim to enhance construction safety, they often fail to adequately consider the needs and rights of workers, who are conscious beings with free will. This issue extends beyond Asimov's first law of robotics, which states that AI systems should not harm humans; it also involves the challenge of programming AI to consistently prioritize human well-being (McAleenan 2020). Sekiguchi and Hori (2020) argue that AI ethics has been neglected in research due to its complexity and the vast amount of information, making it difficult for engineers to relate ethics to their practical work. However, efforts by organizations like the IEEE to develop ethical standards for AI are beginning to bridge this gap, guiding the integration of ethical considerations into engineering practices (Sekiguchi and Hori 2021).

Figure 6.1 summarizes all the necessary procedures and aspects to consider to design, develop, validate, deploy and govern an ethical AI system.

## 6.6  Conclusion

In conclusion, the construction industry is undergoing a profound transformation as it adopts digital technologies under the framework of Industry 4.0 and transitions towards the more human-centric approach of Industry 5.0. The evolution from Construction 4.0 to Construction 5.0 is marked by a shift in focus from automation and efficiency to the integration of human values, sustainability, and ethical considerations in the application of technologies like AI and robotics. While Industry 4.0 technologies have enabled remarkable advancements in productivity, real-time data processing, and decision-making, the introduction of AI into construction also raises

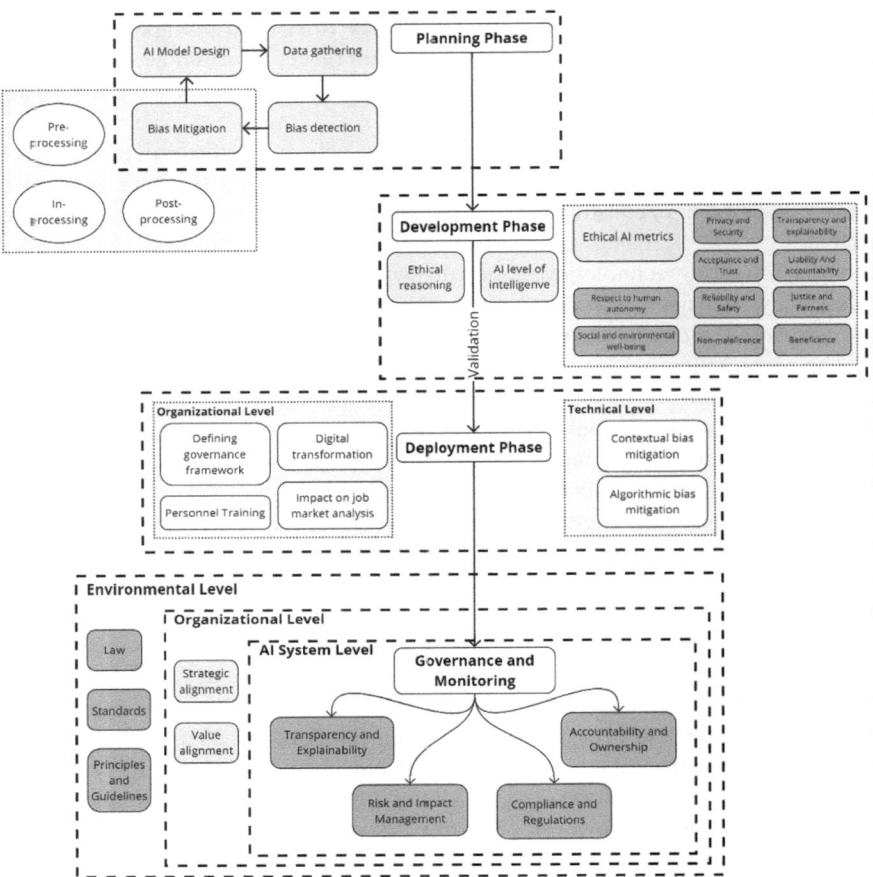

**Fig 6.1** Ethical AI to design, development, deployment, and governance procedures and component (Inspired by (Mäntymäki et al., no date; OECD 2019; Eitel-Porter 2021; Weber-Lewerenz 2021)

significant ethical challenges, particularly around issues such as job displacement, data privacy, and algorithmic biases.

Industry 5.0 offers a potential pathway for addressing these challenges by emphasizing the collaboration between humans and intelligent machines, fostering a more balanced and inclusive technological ecosystem. By prioritizing human well-being, creativity, and ethical norms, Industry 5.0 aims to ensure that AI systems not only improve efficiency but also align with societal values, promoting responsible innovation. The introduction of frameworks such as Responsible AI, ethical AI governance models, and compliance with regulatory standards like the European AI Act provide a foundation for navigating the complex ethical landscape in construction.

As the industry continues to evolve, companies must balance technological advancements with ethical obligations. This requires robust governance systems, bias mitigation strategies, and continuous monitoring to ensure that AI systems enhance human capabilities without undermining trust, fairness, or social equity. Moving forward, the construction sector must focus on workforce training, fostering a culture of ethical awareness, and ensuring that AI applications are transparent, accountable, and aligned with both economic and societal goals. Through these efforts, the construction industry can leverage the full potential of digital technologies while ensuring that the integration of AI is both responsible and beneficial for all stakeholders.

# References

J.G.A.G. Ajay Agrawal, Economic policy for artificial intelligence. AI & Soc. **36**, 101–148 (2019). https://doi.org/10.1007/s00146-020-01043-6. Accessed 8 Sept 2024

S. Akter et al., Algorithmic bias in machine learning-based marketing models. J. Bus. Res. **144**, 201–216 (2022). https://doi.org/10.1016/j.jbusres.2022.01.083

P. Arroyo, A. Schöttle, R. Christensen, The ethical and social dilemma of AI uses in the construction industry, in *IGLC 2021 - 29th Annual Conference of the International Group for Lean Construction - Lean Construction in Crisis Times: Responding to the Post-Pandemic AEC Industry Challenges*. Department of Engineering, Civil Engineering Division, Pontificia Universidad Catolica del Peru (2021), pp. 227–236. https://doi.org/10.24928/2021/0188

A. Askell, M. Brundage, G. Hadfield, The role of cooperation in responsible AI development (2019), http://arxiv.org/abs/1907.04534

B. Balducci, D. Marinova, Unstructured data in marketing. J. Acad. Mark. Sci. **46**(4), 557–590 (2018). https://doi.org/10.1007/s11747-018-0581-x

C. Bartneck et al., Springer briefs in ethics (no date), http://www.springer.com/series/10184

E.L. Deci, A.H. Olafsen, R.M. Ryan, Self-determination theory in work organizations: the state of a science. Guest **4**, 19–43 (2024). https://doi.org/10.1146/annurev-orgpsych, www.annualreviews.org

L.A. Dennis, M. Fisher, A.F.T. Winfield, Towards verifiably ethical robot behaviour (2015), http://arxiv.org/abs/1504.03592

Economist, Artificial intelligence: the impact on jobs – automation and anxiety (2016)

R. Eitel-Porter, Beyond the promise: implementing ethical AI. AI Ethics **1**(1), 73–80 (2021). https://doi.org/10.1007/s43681-020-00011-6

N. Emaminejad et al., Trust in AI and implications for the AEC research: a literature analysis (2015)

N. Emaminejad, R. Akhavian, Trustworthy AI and robotics: implications for the AEC industry. Autom. Constr. **139**, 104298 (2022). https://doi.org/10.1016/j.autcon.2022.104298

European Commission, Digital Services Act (DSA) (2019a)

European Commission, Ethics guidelines for trustworthy AI (2019b)

European Commission, Assessment List for Trustworthy Artificial Intelligence (ALTAI) for self-assessment (2020). https://doi.org/10.2759/791819

European Parliament, Artificial Intelligence Act (2019)

J. Fjeld et al., Principled artificial intelligence: mapping consensus in ethical and rights-based approaches to principles for AI. SSRN Electron. J. [Preprint] (2020). https://doi.org/10.2139/ssrn.3518482

L. Floridi et al., AI4People—an ethical framework for a good AIsociety: opportunities, risks, principles, and recommendations. Mind. Mach. **28**(4), 689–707 (2018)

Future of Life Institute, Asilomar AI principles (2017). https://futureoflife.org/ai-principles/?cnrelo aded=1. Accessed 8 Aug 2024

Gartner, Gartner says nearly half of CIOs are planning to deploy artificial intelligence (2018)

IAPP (International Association of Privacy Professionals), Building ethics into privacy frameworks for big data and AI (2018)

IEEE (Institute of Electrical and Electronics Engineers), Ethically aligned design (2019)

International Organization for Standardization, ISO 9001:2021 quality management systems – requirements. Standard (2021)

Z. Irani, M.M. Kamal, Intelligent systems research in the construction industry. Expert Syst. Appl. **41**(4, Part 1), 934–950 (2014). https://doi.org/10.1016/j.eswa.2013.06.061

A. Kanioura, F. Lucini, Ready. Set. Scale (2019)

A. Khodabakhshian, Ethics-aware application of digital technologies in the construction industry, in *Improving Technology Through Ethics*, eds. by S. Chiodo, D. Kaiser, J. Shah, P. Volonté (Springer Nature Switzerland, Cham, 2024), pp. 49–64. https://doi.org/10.1007/978-3-031-529 62-7_5

A. Khodabakhshian, T. Puolitaival, L. Kestle, Deterministic and probabilistic risk management approaches in construction projects: a systematic literature review and comparative analysis. Buildings **13**(5), 1312 (2023). https://doi.org/10.3390/buildings13051312

B. Kuipers, Perspectives on ethics of AI, in *The Oxford Handbook of Ethics of AI* (2020), https://api.semanticscholar.org/CorpusID:211094766

D. Leslie, Understanding artificial intelligence ethics and safety (2019). https://doi.org/10.5281/zen odo.3240529

C.-J. Liang et al., *Ethics of Artificial Intelligence and Robotics in the Architecture, Engineering, and Construction Industry* (2023)

H. Lu et al., Trustworthy AI: a computational perspective. ACM Trans. Intell. Syst. Technol. **14**(1), 1–59 (2022). https://doi.org/10.1145/3546872

M. Lorenzini et al., Ergonomic human-robot collaboration in industry: a review. Front. Robot. AI. Frontiers Media S.A. (2023). https://doi.org/10.3389/frobt.2022.813907

G. Lorenzoni et al., Machine learning model development from a software engineering perspective: a systematic literature review (2021). http://arxiv.org/abs/2102.07574

P.K.R. Maddikunta et al., Industry 5.0: a survey on enabling technologies and potential applications. J Ind. Inf. Integr. **26**, 100257 (2021), https://api.semanticscholar.org/CorpusID:238648268

D.Ø. Madsen, K. Slåtten, Comparing the evolutionary trajectories of industry 4.0 and 5.0: a management fashion perspective. Appl. Syst. Innov. MDPI (2023). https://doi.org/10.3390/asi 6020048

M. Mäntymäki et al., Putting AI ethics into practice: the hourglass model of organizational AI governance (no date). https://doi.org/10.48550/arXiv.2206.00335

M. Marinelli, From industry 4.0 to construction 5.0: exploring the path towards human–robot collaboration in construction. Systems. MDPI (2023). https://doi.org/10.3390/systems11030152

Marriam-Webster Dictionary (no date)

R. Maskuriy et al., Industry 4.0 for the construction industry: review of management perspective. Economies. MDPI Multidisciplinary Digital Publishing Institute (2019). https://doi.org/10.3390/economies7030068

P. McAleenan, Moral responsibility and action in the use of artificial intelligence in construction. Proc. Inst. Civ. Eng.: Manag. Procure. Law **173**(4), 166–174 (2020). https://doi.org/10.1680/jmapl.19.00056

McKinsey Global Institute, Bridging global infrastructure gaps (2016), www.mckinsey.com/mgi

Mckinsey Global Institute, Jobs lost, jobs gained: workforce transitions in a time of automation (2017), www.mckinsey.com/mgi

I. Mezgár, J. Váncza, From ethics to standards – a path via responsible AI to cyber-physical production systems. Annu. Rev. Control. Elsevier Ltd. 391–404 (2022). https://doi.org/10.1016/j.arc ontrol.2022.04.002

D.F. Mujtaba, N.R. Mahapatra, Ethical considerations in AI-based recruitment, in *2019 IEEE International Symposium on Technology and Society (ISTAS)* (2019), pp. 1–7, https://api.semantics cholar.org/CorpusID:209459647

V.C. Müller, Ethics of artificial intelligence and robotics (no date), http://plato.stanford.edu/

M.A. Musarat et al., Digital transformation of the construction industry: a review, in *2021 International Conference on Decision Aid Sciences and Application, DASA 2021* (Institute of Electrical and Electronics Engineers Inc., 2021), pp. 897–902. https://doi.org/10.1109/DASA53625.2021.9682303

S. Nahavandi, Industry 5.0-a human-centric solution. Sustainability (Switzerland) 11(16), 4371 (2019). https://doi.org/10.3390/su11164371

T. Nam, Technology use and work-life balance. Appl. Res. Qual. Life 9(4), 1017–1040 (2014). https://doi.org/10.1007/s11482-013-9283-1

National Institute of Standards and Technology (NIST), Cybersecurity framework: A starter guide (2021)

OECD, *Scoping the OECD AI principles: deliberations of the expert group on artificial intelligence at the OECD (AIGO)* (2019). https://doi.org/10.1787/d62f618a-en

G. Orrù et al., Human-like problem-solving abilities in large language models using ChatGPT. Front. Artif. Intell. 6, 1199350 (2023). https://doi.org/10.3389/frai.2023.1199350

V.S. Pillai, K.J.M. Matus, Towards a responsible integration of artificial intelligence technology in the construction sector. Sci. Public Policy 47(5), 689–704 (2020). https://doi.org/10.1093/sci pol/scaa073

S.A. Prieto, E.T. Mengiste, B. García de Soto, Investigating the use of ChatGPT for the scheduling of construction projects. Buildings 13(4), 857 (2023). https://doi.org/10.3390/buildings13040857

A. Rai, S. Sarker, EDITOR'S COMMENTS next-generation digital platforms: toward human-AI hybrids. MIS Q. (2019), https://android-developers.googleblog.com/2018/01/how-we-fought-bad-apps-and-malicious.html

M. Regona et al., Opportunities and adoption challenges of AI in the construction industry: a PRISMA review. J. Open Innov.: Technol. Mark. Complex. 8(1), 45 (2022). https://doi.org/10.3390/joitmc8010045

D. Rozado, Wide range screening of algorithmic bias in word embedding models using large sentiment lexicons reveals underreported bias types. PLOS ONE 15(4), 1–26 (2020). https://doi.org/10.1371/journal.pone.0231189

A. Santhosh et al., AI impact on job automation. Int. J. Eng. Technol. Manag. Sci. 7(4), 410–425 (2023). https://doi.org/10.46647/ijetms.2023.v07i04.055

A. Schöttle, What drives our project teams? in *Proceedings of the 28th Annual Conference of the International Group for Lean Construction (IGLC)*. Berkeley, California, USA (2020)

K. Sekiguchi, K. Hori, Designing ethical artifacts has resulted in creative design. AI & Soc. 36(1), 101–148 (2021). https://doi.org/10.1007/s00146-020-01043-6

K. Siau, W. Wang, Artificial intelligence (AI) ethics: ethics of AI and ethical AI. J. Datab. Manag. IGI Global 74–87 (2020). https://doi.org/10.4018/JDM.2020040105

J. Svegliato, S.B. Nashed, S. Zilberstein, An integrated approach to moral autonomous systems, in *European Conference on Artificial Intelligence* (2020), https://api.semanticscholar.org/Cor pusID:211249836

N. Syam, A. Sharma, Waiting for a sales renaissance in the fourth industrial revolution: Machine learning and artificial intelligence in sales research and practice. Ind. Mark. Manag. 69, 135–146 (2018). https://doi.org/10.1016/j.indmarman.2017.12.019

The Public Voice, Universal guidelines for artificial intelligence (2018)

S. Thiebes, S. Lins, A. Sunyaev, Trustworthy artificial intelligence. Electron. Mark. 31(2), 447–464 (2021). https://doi.org/10.1007/s12525-020-00441-4

J. Torresen, A review of future and ethical perspectives of robotics and AI. Front. Robot. AI. Frontiers Media S.A. (2018). https://doi.org/10.3389/frobt.2017.00075

P. Tunji-Olayeni et al., Research trends in industry 5.0 and its application in the construction industry. Technol. Sustain. 3(1), 1–23 (2024). https://doi.org/10.1108/TECHS-07-2023-0029

UNESCO, Report of world commission on the ethics of scientific knowledge and technology on robotics ethics (2017)

W. Wallach, C. Allen, *Moral Machines: Teaching Robots Right from Wrong* (Oxford University Press, Oxford, 2009). https://doi.org/10.1093/acprof:oso/9780195374049.001.0001

C.G. Walsh et al., Stigma, biomarkers, and algorithmic bias: recommendations for precision behavioral health with artificial intelligence. JAMIA Open. Oxford University Press, 9–15 (2021). https://doi.org/10.1093/JAMIAOPEN/OOZ054

W. Wang, K. Siau, Artificial intelligence, machine learning, automation, robotics, future of work and future of humanity: a review and research agenda. J. Datab. Manag. **30**(1), 61–79 (2019). https://doi.org/10.4018/JDM.2019010104

B. Weber-Lewerenz, Corporate digital responsibility (CDR) in construction engineering—ethical guidelines for the application of digital transformation and artificial intelligence (AI) in user practice. SN Appl. Sci. **3**(10), 1–25 (2021). https://doi.org/10.1007/s42452-021-04776-1

D. Weinberger, Don't make AI artificially stupid in the name of transparency. Wired (2018)

A.F.T. Winfield, C. Blum, W. Liu, Towards an ethical robot: internal models, consequences and ethical action selection, in *Advances in Autonomous Robotics Systems*. ed. by M. Mistry et al. (Springer International Publishing, Cham, 2014), pp.85–96

A.F.T. Winfield, M. Jirotka, Ethical governance is essential to building trust in robotics and artificial intelligence systems. Philos. Trans. r. Soc. A Math. Phys. Eng. Sci. **376**(2133), 20180085 (2018). https://doi.org/10.1098/rsta.2018.0085

W. Wu, T. Huang, K. Gong, Ethical principles and governance technology development of AI in China. Engineering. Elsevier Ltd. 302–309 (2020). https://doi.org/10.1016/j.eng.2019.12.015

M.C. Zizic et al., From Industry 4.0 towards industry 5.0: a review and analysis of paradigm shift for the people, organization and technology. Energies. MDPI (2022). https://doi.org/10.3390/en15145221